The Shell of a Typical Prosobranch, or Gill-breathing Snail

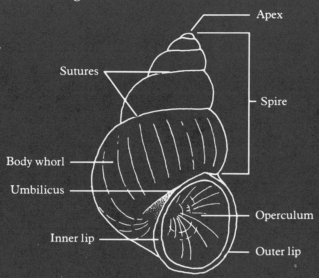

The Shell of a Typical Pulmonate, or Lung-breathing Snail

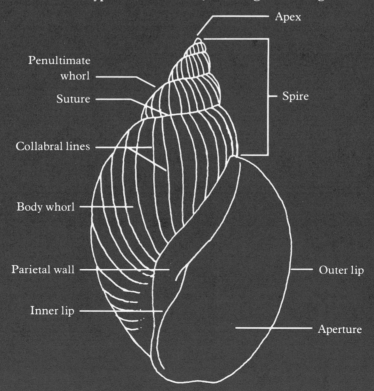

21374
17.50

The Freshwater Molluscs of Canada

A female Pocket-Book mussel (*Lampsilis ventricosa*) with its mantle flap extended and pulsating to simulate a wounded minnow. This activity attracts fish, and increases the opportunities for juvenile mussels (glochidia) to attach themselves to a fish after they are ejected from the parent. A period of attachment to a fish of the proper species is necessary for maturation of most juvenile freshwater mussels. (Photograph courtesy of John H. Welsh, Boothbay, Maine.)

The Freshwater Molluscs of Canada

Arthur H. Clarke

National Museum of Natural Sciences
National Museums of Canada

© National Museums of Canada 1981

Published by the
National Museum of Natural Sciences
National Museums of Canada
Ottawa, Canada K1A 0M8

Catalogue No. NM95-17/5

Printed in Canada

ISBN 0-660-00022-9

Édition française
Les Mollusques d'eau douce du Canada
ISBN 0-660-00023-7

Managing editor: Viviane Appleton
Text editor: Lorraine Smith
Production: Donald Matheson
           James MacLeod
Design: Eiko Emori
Typesetting: The Runge Press Ltd.
Printing: D. W. Friesen & Sons Ltd.

Contents

Acknowledgements 9

Introduction 11
    Why Collect Shells? 11
    How to Collect Freshwater Shells 12
    Arrangement and Care of the Collection 16
    Scientific Names 17
    Mollusc Classification and Some Special Features 19
    Distribution Patterns in Canada 21
    Molluscs as Pollution Indicators 24

The Freshwater Molluscs 27

Key to the Families of Canadian Freshwater Molluscs 29
Class Gastropoda (Snails) 31
    Subclass Prosobranchia (Gill-breathing Snails) 33
    Order Mesogastropoda 33
     I Superfamily Viviparacea 33
       Family Viviparidae (Mystery Snails) 33
     II Superfamily Valvatacea 41
       Family Valvatidae (Valve Snails) 41
     III Superfamily Rissoacea 55
       Family Hydrobiidae (Spire Snails) 55
       Family Truncatellidae (Looping Snails) 73
       Family Bithyniidae (Faucet Snails) 77
     IV Superfamily Cerithiacea 81
       Family Pleuroceridae (Horn Snails) 81
    Subclass Pulmonata (Lung-breathing Snails) 89
    Order Basommatophora 89
     V Superfamily Acroloxacea 89
       Family Acroloxidae (Primitive Freshwater Limpets) 89
     VI Superfamily Lymnaeacea 93
       Family Lancidae (Limpet-like Lymnaeas) 93
       Family Lymnaeidae (Pond Snails) 97

   VII Superfamily Physacea   151
         Family Physidae (Tadpole Snails)   151
  VIII Superfamily Planorbacea   175
         Family Planorbidae (Ramshorn Snails)   175
         Family Ancylidae (True Freshwater Limpets)   219

Colour Plates   229

Class Pelecypoda (Clams and Mussels)   245
  Order Eulamellibranchia   247
   IX Superfamily Unionacea (Freshwater Mussels)   247
        Family Margaritiferidae (Pearly River-Mussels)   247
        Family Unionidae (Pearly Mussels)   253
          Subfamily Ambleminae (Button Shells and Relatives)   254
          Subfamily Anodontinae (Floater Mussels)   272
          Subfamily Lampsilinae (Lamp Mussels)   312
    X Superfamily Sphaeriacea   357
        Family Corbiculidae (Little Basket Clams)   357
        Family Sphaeriidae (Fingernail Clams and Pea Clams)   361
          Subfamily Sphaeriinae (Fingernail Clams)   362
          Subfamily Pisidiinae (Pea Clams or Pill Clams)   386

Glossary   433

References   437

Index to Scientific and Common Names   441

Acknowledgements

I am grateful to Arthur R. Clarke, the late Louise R. Clarke, the late Françoise Dehenne, F. Wayne Grimm, Brian T. Kidd, Judith J. McDonald and the late Dr. D. G. S. Wright for field assistance; to Walter MacKay Drycott, the late Reverend H. B. Herrington and several other colleagues for contributing some valuable specimens; and to Muriel F. I. Smith and Jane M. Topping for laboratory assistance.
Dr. G. L. Mackie gave constructive criticism on the section on Sphaeriidae.

I am also indebted to Dr. Aurèle LaRocque, who, through energetic application of his expert knowledge of molluscs and of the French language, provided the fine translation for the French edition of this book.

The watercolour paintings are by Valerie Fulford (Figures 110, 128, 129, 131, 134-38, 140, 142, and 143), Jacques Blais (Figure 139), and Aleta Karstad Schueler (all others). Photographs are by the Canadian Conservation Institute of the National Museums of Canada (most of the *Pisidium*, using a scanning electron microscope), the National Museum of Natural Sciences (Unionacea and *Corbicula*), and A. M. Frias-Martins (all others). My work was supported by the National Museum of Natural Sciences and by the Arctic Biological Station, Fisheries and Oceans Canada.

Introduction

Why Collect Shells

Canadians are exceedingly fortunate in having an abundance of beautiful unspoiled forests, lakes, and rivers. Thousands have found that frequent escape to these open spaces for fresh air, peace, and relaxation is essential for their mental and physical well-being.

As with music and art, knowledge enhances the appreciation of nature. Handbooks for the identification of trees, flowers, mushrooms, mammals, birds, fishes, and insects are justifiably popular. Interest in seashells, freshwater shells and land snails is also widespread. In fact, many people who wish to make natural-history collections find mollusc shells more practical and desirable than anything else.

There are many reasons why people collect shells. Shells are beautiful, intriguing, and free for the taking, and collecting them is fun. Shells are easy to find, and unlike plants and other animals require very little special preparation or preservation. In addition to healthful exercise, shell collecting provides an interesting diversion in unfamiliar regions. Shells make excellent souvenirs because, as faunal samples, they represent the very essence of any exotic region.

Shell collecting often brings about good companionship with other collectors. It many also lead to membership in local shell clubs or in national and international organizations. Enthusiastic shell collectors live in all parts of the world, and a great many of them are eager to exchange their exotic native shells for our ordinary ones, which to them are just as exotic.

As familiarity with molluscs grows, most collectors discover that living molluscs are fascinating creatures to observe. Original and important contributions to science can be made in this way. Many molluscs are beautiful and colourful, some can ascend and descend through the water apparently at will, others have soft parts that look like and pulsate like an injured minnow (see frontispiece), and all have interesting feeding and reproductive habits. A great many basic facts about our most common species are still unknown. For example, during which times of the year do they reproduce in different regions? What do they eat? What are their major predators? And what are their limits of distribution? It is still possible, in fact, to discover new species that are entirely unknown to science.

A shell collector's interest might originally be aroused by seashells. Later, freshwater shells and land snails might be added to the collection, especially by someone who lives in an inland location. Many collectors then decide to specialize in freshwater or land shells. Several popular handbooks exist for the identification of seashells, but non-technical books on freshwater or land molluscs are very scarce, and comprehensive texts covering all of Canada do not exist.

This book is about the freshwater shells of Canada. It attempts to describe and illustrate all the species that live in this country, to provide keys and descriptions for their identification, and to present information about their ecology and their relationship to man.

How to Collect Freshwater Shells

Freshwater molluscs are abundant animals and are usually not difficult to collect. The successful collector, however, must learn to recognize a likely habitat.

In general, well-vegetated portions of unpolluted lakes, ponds, and slow-flowing rivers are the most productive localities for freshwater snails (class Gastropoda), which occur there on submersed vegetation, on rocks, and on the bottom from the water's edge out to a considerable depth. Freshwater mussels (family Unionidae) are found principally in rivers and lakes, partly buried in the bottom, where the water is about one-third of a metre to two metres deep. The greatest diversity of species occurs in rather fast-flowing portions of hard-water rivers, that is those containing a moderate or high concentration of dissolved limestone, such as the rivers of southern Ontario and southern Manitoba. Fingernail clams and pill clams (family Sphaeriidae) are most abundant in the muddy and fine sandy bottoms of lakes. They burrow just below the bottom in water a few centimetres to many metres deep.

Several important exceptions to these general rules exist. Temporary springtime, that is vernal, pools often support thriving colonies of snails, especially *Aplexa hypnorum* and *Gyraulus circumstriatus* in the east and *Planorbula campestris* and *Stagnicola caperata* in the west. Tiny spring-fed pools in the west may contain *S. montanensis*. Wave-beaten rocky shores of large lakes may harbour the rare limpet *Acroloxus coloradensis* on the undersurfaces of boulders. Arctic muskeg pools often teem with *Physa jennessi jennessi* and *S. arctica*. Brackish-water estuaries may support rich freshwater-mollusc populations in very shallow water above the level of the lower saltwater layer. Deep water in large lakes may harbour *Fossaria decampi* and species of *Valvata* and *Pisidium*. Rocky creeks in the far east may contain the pearly freshwater mussel *Margaritifera margaritifera*, and those in the far west the related species *M. falcata*.

Freshwater molluscs can be collected by hand and put in a pocket. But in most cases hand-picking is slow, and pocketed specimens are likely to be crushed. Therefore, although not essential, a certain amount of equipment is very useful.

The usual gear used in collecting freshwater molluscs includes
1) a pair of hip-waders or, in warm weather, tennis shoes and a bathing suit
2) a dip net for sweeping underwater plants and scraping the bottom
3) a glass-bottomed box or bucket for observing the bottom (used mainly when collecting freshwater mussels)
4) a supply of suitable jars and vials with leakproof caps
5) 80 per cent ethyl alcohol or 70 per cent isopropyl alcohol (70 per cent ordinary rubbing alcohol and 30 per cent water) for preserving the catch
6) cloth bags with string or cloth-strip closures for freshwater mussels
7) field labels made of plastic or durable paper (that is, with high rag content)
8) a field notebook
9) pencils, or a pen and waterproof ink
10) forceps
11) a flat pan for sorting small specimens
12) detailed maps of the region to be explored.

Additional useful gear includes Nembutal or propylene phenoxetol for relaxing and extending specimens, 10 per cent formalin buffered with borax for initial preservation of specimens (but *not* for long-term storage), rope and a small dredge for collecting in deep water, a rowboat, outboard motor, oars and life-jackets, an underwater face-mask, fins and snorkel, and Scuba gear.

A widely used collecting technique is as follows: the collector wades into a river and walks slowly upstream, carefully scanning the bottom through the glass-bottomed bucket. A sunny day is best for visibility, and the lower the water level the better. By working upstream, the collector will find that disturbed mud drifts behind, leaving the water in front clear. Partly buried freshwater mussels will be obvious, and fully buried ones will reveal themselves as white slits in the river-bottom. These are the mantle edges as seen between the slightly gaping valves. The glass-bottomed bucket, which should be tied to the collector's waist, is a convenient receptacle for mussels.

Once the search for mussels has been completed and the specimens have been placed in cloth bags, the glass-bottomed bucket can be set aside. The dip net is now used for collecting snails from vegetation and various small molluscs from the river bottom. Rocks should also be picked up, examined all over for snails, and then replaced in their original position. Restoring the habitat as closely as possible to its original state is an important aspect of responsible and conservation-oriented collecting. Live specimens should be put into jars with enough water to keep them alive. It takes at least 30 minutes to search a productive locality efficiently. Some collectors, in fact, will spend four to six hours at a collecting site to ensure that the rare species, as well as the common ones, have been found.

If snails and sphaeriids are to be kept alive, they should be cool and uncrowded, that is in water that is about ten times greater in volume than the specimens themselves and in a jar large enough to provide half that much air volume. Mussels will remain alive for at least a day if placed in cloth bags kept out of the sun and moistened every few hours. They should not be kept crowded in a bucket of water, because they will soon deplete the dissolved oxygen and die.

If specimens are to be of any interest to serious collectors or of any value to science, accurate records must be kept. Minimal data should include
1) the name of the water body the specimens were found in
2) the distance and direction from the centre of the nearest city, town, or village
3) the province, or state, and country
4) the date
5) the collector's name.
Accessory valuable data are
1) the depth of the water where the specimens were collected
2) the kind of bottom (mud, sand, boulders, for example)
3) the kind and density of vegetation
4) the size (approximate width or area) of the water body
5) the approximate current speed
6) the water temperature
7) any other relevant observations

Collection sites are referred to as stations, and these are best numbered consecutively in a continuous series. The station number and usually the minimal data are written on a field label placed in the container with the specimens. Jars of living snails should also have the station number written on the cap, because the snails may eat the label. The same number and the minimal and accessory data are recorded in a field notebook. On an extended trip, carbon copies of field notes should be mailed home periodically. A lost field-book can ruin a collecting trip.

Many collectors will wish to remove and discard the soft parts of their specimens soon after collection, before the animals die and begin to decompose. Freshwater mussels and fingernail clams are usually cleaned by boiling. This causes the valves to gape, and the cooked, firm body of the clam can then be easily removed. The soft parts of large snails can also be loosened by boiling, and then "unscrewed" from their shells. A bent pin is useful for this. Small snails and pill clams can be spread out on a flat pan or a newspaper and simply allowed to dry out. Soaking in preserving fluid prior to drying is helpful, but is not necessary if an open-air drying area is available and if only a few small molluscs are involved.

Many collectors might be interested to know that freshwater mussels from unpolluted habitats are edible. They may be steamed, roasted, fried, or made into chowder. The reader is urged to experiment. If the water is clean enough to drink, the mussels are clean enough to eat.

A note on conservation is necessary here. Some of our freshwater species are, or are becoming, very rare because of habitat disruption, increased urbanization, pollution, or over-collecting. Certain freshwater mussels, especially *Simpsoniconcha ambigua* and species of *Dysnomia*, are particularly threatened. Collectors are urged to be conservation minded—to collect only as many live specimens as they need, and never to collect all available specimens of any species in a particular locality. The responsible collector also avoids disrupting habitats. Live collecting is not always necessary because empty shells often make perfectly good specimens for the collection. Beach-drift specimens or empty shells left by muskrats are often in excellent condition.

For more extensive information on collecting see *How to Collect and Study Shells*, published by the American Malacological Union and available for US $2.50 from the Department of Malacology, Academy of Natural Sciences of Philadelphia, 19th and the Parkway, Philadelphia, Pa. 19103, USA. Another useful publication, available from the same source for US $1.50, is entitled *Papers on Rare and Endangered Mollusks of North America*, and was originally published in the scientific journal *Malacologia* (vol. 10, no. 1, 1970).

Arrangement and Care of the Collection

After they are cleaned or preserved, all the specimens from a particular collecting station should be sorted into vials or boxes by species; that is, each species should have its own container. A slip of paper bearing the station number and the species or subspecies name should accompany each such lot. Each lot is then catalogued as follows:

1) a catalogue number is assigned and written in india ink directly on the shell or on a small label that is placed with the shells inside each closed container

2) that same catalogue number is written on a larger label, along with the name of the species or subspecies and author,

all minimal data and, if space permits, any accessory data, and is placed in the shallow cardboard boxes described below
3) the same information is entered in a ledger book (the catalogue).
In this way, the data associated with each species lot will be protected from loss, and the lasting value of the collection will be assured.

Museums use dustproof cabinets with large, shallow wooden drawers to house mollusc collections. Shallow cardboard boxes of various multiple sizes (say $5 \times 3.5 \times 2$ cm deep, $5 \times 7 \times 2$ cm, $10 \times 7 \times 2$ cm, etc.) are placed in the drawers, and all lots of the same species are placed together in horizontal or vertical rows. The species are arranged together with other species of the same genus, all genera in the same family being stored together, and so on. These are all arranged in systematic, or phylogenetic, order, that is in their presumed evolutionary sequence, from the most primitive to the most advanced groups. The drawers are also labelled to indicate the genera or species they contain. The arrangement of species in this book follows the phylogenetic order advocated by Taylor and Sohl (1962) and Moore (1969).

In practice, many collectors are obliged to use miscellaneous vials and boxes. Cigarette boxes make good containers for small shells, and may be stored in shoe boxes on shelves. Many kinds of containers will do, but of course uniformity enhances the appearance of the collection.

Scientific Names

Systematics has been defined as the science of deducing evolution in action. Through classification and the use of names, it seeks to show evolutionary relationships among animals and plants.

Living creatures are usually divided into two major groups, the Animal Kingdom and the Plant Kingdom. The major divisions of the Animal Kingdom are called phyla (Chordata, Arthropoda, Mollusca, for example). Phyla are further subdivided into classes, classes into orders, orders into families, families into genera, genera into species, and sometimes species into subspecies. Occasionally, to show relationships more precisely, it is necessary to use additional categories, for example subclass between class and order, and subgenus between genus and species.

A species includes all individuals potentially able to breed together to produce similar individuals. Some species have common names, but these may be local and not known to people who live elsewhere or speak another language. The scientific name of an animal is understood internationally and is vastly preferable to a confusing variety of local common names.

The scientific name of a species consists of the genus name, with the first letter capitalized; the subgenus name, if applicable, in round brackets, also with the first letter capitalized; and the specific name, uncapitalized. It is usual also to include the name of the author who first described and named the animal and the date of publication. Following approved practice, if the specific name used for the animal in this book is now in a different genus from the one originally proposed, the original author and date are placed in round brackets. An example is *Stagnicola (Hinkleyia) montanensis* (Baker, 1913). This species was first described and named by F. C. Baker in 1913. However, Baker named it *Galba montanensis*.

A subspecies is a population, or group of populations, that is significantly different from other populations of that species but is still potentially able to breed with those populations. Subspecies of the same species also occupy different areas. If these areas overlap, widespread interbreeding can be expected in that zone. A subspecies name consists of the complete species name followed by the subspecific name. The same rules concerning authors, dates and use of brackets apply. An example is *Stagnicola (Stagnicola) catascopium preblei* (Dall, 1905). This subspecies was described by W. H. Dall in 1905 as the species *Lymnaea preblei*.

The interested reader is referred to Ross (1974) and Mayr (1969), and to the scientific journal *Systematic Zoology*, for fuller discussion of classification procedure and the formation of scientific names.

Molluscan Classification and Some Special Features

Snails, slugs, clams, mussels, scallops, oysters, chitons, tooth shells, squids and octopuses are all molluscs. The features these animals have in common constitute a good definition of what molluscs are.

The phylum Mollusca may also be defined as comprising all invertebrate animals that are soft-bodied, non-segmented, have a muscular foot for burrowing or crawling, and possess a mantle—an enveloping sheet of tissue that in most species secretes a calcareous shell. With the major exception of pelecypods (bivalves), most molluscs also have a head with tentacles, eyes, mouth, and a radula—a rasping organ in the floor of the mouth.

Seven classes of living molluscs are usually recognized: Monoplacophora, Polyplacophora, Aplacophora, Scaphopoda, Pelecypoda, Gastropoda, and Cephalopoda. Only the two largest classes, Pelecypoda and Gastropoda, live in fresh water, and only Gastropoda live on the land. All of the classes, however, are found in the sea. General information on molluscan classes is given in most textbooks on general biology. For detailed information on marine groups, however, consult Abbott (1974) or Bousfield (1960), and on land snails Pilsbry (1939-48) or Burch (1962).

Canadian freshwater gastropods belong to two subclasses—Prosobranchia (also called Streptoneura) and Pulmonata (which, together with Opisthobranchia, are also called Euthyneura). Most prosobranchs have an operculum (used to seal the shell aperture), breathe by means of gills, and individuals are either male or female. Pulmonates have no operculum, breathe by means of a pulmonary sac, or lung, and are hermaphroditic. Canada has four superfamilies of freshwater prosobranchs (divided into six families) and four superfamilies of pulmonates (also containing six families).

Most systems of classification do not subdivide the class Pelecypoda into subclasses but only into about five orders. All Canadian freshwater bivalves belong to the order Eulamellibranchia. This group is characterized by
1) a hinge containing a few teeth of diverse shapes and sizes
2) two large adductor muscles of about the same size, one anterior and one posterior

3) a partly closed mantle with well-developed siphons
4) leaf-like gills within the mantle cavity.
Two superfamilies are represented in Canada—Sphaeriacea and Unionacea.

The major external and internal features of prosobranch and pulmonate snails and of eulamellibranch bivalves are illustrated on the front and back endpapers.

Snails have many unique features, but their feeding organ—the radula—is one of the most interesting. This is a tough, elongate, moveable membrane in the mouth that bears many similar transverse rows of tiny sharp teeth. The radula is pressed against the food and pulled back and forth. This action rasps the food into tiny particles and carries it back into the mouth.

Scientists have shown that the number and shape of the radular teeth in each transverse row are very useful in classification. All Canadian prosobranch snails, for example, typically have 7 teeth in each transverse row. The formula for this is usually 2-1-1-1-2, meaning that each row has 2 marginal teeth, 1 lateral, 1 central, another lateral, and 2 more marginals, in that order. The shape of the teeth and the number of projections on each (the "cusps") differs from family to family, but is generally similar within families and very similar within genera. Similarly, the pulmonates are characterized by having radular teeth set in rows, with one distinctive central tooth in each row flanked by many teeth on either side. These teeth may or may not be separable into laterals and marginals. Each family of pulmonates has very distinctive radulae. For detailed information on radulae see the fascinating book by Solem (1974).

Another uniquely molluscan feature is the glochidium larva of freshwater mussels. These animals retain their young for various lengths of time in modified portions of the gills. The young mussels, the glochidia, are released by the parent when its light-sensitive mantle-spots are stimulated, for example by the shadow of a passing fish. Mussels of the genus *Lampsilis* and their relatives even possess special mantle structures apparently designed to lure fish into their vicinity. The glochidia of each species of mussel, with a few exceptions, must attach to the gills or fins of a fish belonging to one, or a few, species for further development to take place. Most glochidia never accomplish this, but those that do succeed remain attached for a few weeks and metamorphose into tiny mussels. They then drop to the bottom and take up the normal life of a mussel, that is they crawl around siphoning water for respiration and to obtain phytoplankton as a source of nourishment and growth.

Distribution Patterns in Canada

Every region of Canada has its own species of freshwater molluscs. For example, that part of southern Ontario which includes Lake Erie and Lake St. Clair and the rivers and streams draining into them harbours twelve species of freshwater mussels and one species of pill clam that occur nowhere else in Canada. Such an area is called a zoogeographic region. Several such regions exist in Canada (see front endpaper).

The most clearly defined regions, that is those in which many species have similar distributional limits, are the Lake Erie-Lake St. Clair Region, the Red River-Assiniboine River Region, and to a lesser extent the Pacific Coastal Region. The other regions have boundaries that are not sharply defined. A general knowledge of the zoogeographical regions of Canada will nevertheless be very useful to a collector.

There are many interesting reasons why Canadian species are distributed as they are, and why Canadian zoogeographic regions exist as they do. These reasons involve the glacial and postglacial history of the country, its climate, geology, and geography, and the biology of the species themselves. For more details than those that follow, see Clarke (1973).

During four different periods of the Pleistocene epoch, which itself lasted from about one million to about five thousand years ago, most of Canada was covered with glacial ice. For freshwater molluscs the most important non-glaciated regions were a part of the Yukon Territory and Alaska known as the Beringian Refugium, and the large area south of the glacial ice, most of which is in the United States. Within the glaciated region all the freshwater molluscs were wiped out, of course, but each time the ice receded they reinvaded the previously ice-covered region.

Many species, especially most of the freshwater mussels and all of the large prosobranch snails (Viviparidae and Pleuroceridae), require continuous waterways for migration. Glochidia of freshwater mussels may be carried over long distances while attached to their fish hosts. The present distributions of freshwater mussels and large operculate snails were therefore brought about largely by postglacial stream confluences, for example the drainage pathways taken by glacial meltwater. Such species are most useful for delimiting zoogeographic regions. On the other hand, most small snails and many small clams (sphaeriids) are probably carried about imbedded in the feathers of water birds or in mud attached to their feet. Sphaeriids may also be transported while clamped to the feet of large aquatic flying insects. Therefore, the present distributions of small molluscs tend to transcend zoogeographic and drainage-system boundaries.

The zoogeographic regions based on the occurrence of freshwater molluscs in Canada are as follows:

1) *The Atlantic Coastal Region*

    This area was populated by freshwater molluscs chiefly from the Atlantic Coastal Plain to the south. Characteristic species and subspecies are *Lyogyrus granum, Margaritifera margaritifera, Anodonta cataracta cataracta, A. implicata,* and *Lampsilis ochracea.*

2) *The Lake Erie-Lake St. Clair Region*

    This rich drainage area in south-central Ontario was populated from the Ohio-Mississippi system when glacial meltwater in that region flowed south. It has twelve unionid species not found elsewhere in Canada, including *Quadrula pustulosa, Cyclonaias tuberculata, Pleurobema coccineum, Ptychobranchus fasciolaris,* and *Obliquaria reflexa.* Some of its species also extend into Region 3.

3) *The Great Lakes-St. Lawrence Region*

    The mollusc fauna of this area was derived from both the Atlantic Coastal Plain and the American Interior Basin. Characteristic species are *Valvata perdepressa, Pleurocera acuta, Goniobasis livescens, Acella haldemani,* and *Alasmidonta marginata.*

4) *The Red River-Assiniboine River Region*
This rich area was populated from the upper Mississippi River by migration through glacial meltwater channels and possibly through recent confluence. Characteristic species, many of which are found in Regions 2 and 3 but originate from another source, include *Cincinnatia cincinnatiensis*, *Amblema plicata*, *Fusconaia flava*, *Quadrula quadrula*, and *Proptera alata*. Some of the species extend into Region 5.

5) *The Western Prairie Region*
This large region in southern Manitoba, Saskatchewan, and Alberta was populated from the American Interior Basin. Its characteristic species and subspecies are *Bakerilymnaea bulimoides*, *Stagnicola caperata*, *Promenetus exacuous megas*, *Planorbula campestris*, and *Helisoma trivolvis subcrenatum*.

6) *The Pacific Coastal Region*
This region covers most of British Columbia and was populated from the Pacific coastal area in the United States. Characteristic species are *Fossaria truncatula*, *Physa columbiana*, *Margaritifera falcata*, *Gonidea angulata*, and *Anodonta nuttalliana*. As some species occur only in the Columbia River system, that area might be designated as a distinct subregion.

7) *The Beringian Refugium*
In Canada this region covers only the Yukon River system in the Yukon Territory and northern British Columbia and some small river systems in the northwestern Northwest Territories. It was a refuge area during the glacial period for many species, most of which have now spread beyond its borders. Its most distinctive species are *Lymnaea atkaensis*, *Stagnicola kennicotti*, and *Anodonta beringiana*.

8) *The Subarctic Region*
The largest faunal zone in Canada, this region extends south of the tree line from Labrador to the mouth of the Mackenzie River. It largely coincides with the Boreal Forest botanical region, and was populated chiefly from adjacent regions to the south. Characteristic species and subspecies are *Stagnicola catascopium preblei*, *Anodonta grandis simpsoniana*, *Sphaerium nitidum*, and *Pisidium conventus*.

9) *The Arctic Region*

This area extends north of the tree line to the southern part of the Arctic Archipelago, and its fauna was probably derived chiefly from the Beringian Refugium. No freshwater molluscs occur farther north. Characteristic species and subspecies are *Valvata sincera helicoidea*, *Stagnicola arctica*, and *Physa jennessi jennessi*.

Thus, a collector travelling from one zoogeographic region to another will encounter different species. The whole molluscan fauna will not change, however, because most species occur in more than one zoogeographic region. For example, the common freshwater mussel *Elliptio complanata* is found in Regions 1, 3, and 8, the abundant pulmonate snail *Stagnicola elodes* occurs everywhere except in the Arctic Region, and the ubiquitous pea clam *Pisidium casertanum* occurs in all regions.

Molluscs as Pollution Indicators

Mollusc populations are affected by all three kinds of water pollution—thermal, inorganic, and organic.

Thermal pollution sometimes occurs where river water is used for industrial cooling. This may heat the water enough to kill molluscs outright or adversely affect their reproductive cycles. If the warming is minor, however, mollusc populations may actually increase.

Inorganic pollution is principally industrial. It may poison water to the extent of killing all molluscs. Inorganic pollution that kills some species but not others, however, is difficult to detect biologically; no precise tolerance limits are known for any species of freshwater mollusc exposed to particular inorganic pollutants. Freshwater mussels live for many years, and during each winter form a more or less distinct growth ring on their shells. Chemical analysis of the shells can reveal if water pollution from radioactive materials or heavy metals has occurred, and when.

Organic pollution is usually caused by sewage and insecticides. Chemical analysis of the soft parts of molluscs can reveal recent pollution by insecticides. In natural water, sewage is first attacked by bacteria that use oxygen. Because bacteria increase rapidly in the presence of sewage, all the oxygen dissolved in water may be used up. Nevertheless, as the sewage is degraded and if no more sewage is added, a sewage-polluted river will become reoxygenated downstream, and the water will become clean.

In water with little oxygen, pulmonate snails that come to the surface to breathe air have an advantage over other molluscs that must extract dissolved oxygen from water with their gills. Gill-breathing molluscs are therefore the first to be killed by low oxygen concentrations. A few gill-breathers (for example *Amblema plicata*, *Anodonta cataracta cataracta*, *A. grandis grandis*, *Sphaerium transversum*, *S. striatinum*, and *Campeloma decisum*) appear to be reasonably tolerant of low oxygen concentrations. Pulmonates, however, are much more tolerant.

Low diversity, and especially the presence of only a single species (usually of the genus *Physa*), is often indicative of organic pollution. *Physa*, in fact, may be unusually abundant in mildly polluted water because the species of fish that would normally prey upon it cannot tolerate the polluted environment. As a general rule, the presence of highly diverse communities of freshwater molluscs gives assurance of clean water.

No freshwater mussels are found in grossly polluted water. Their absence from mildly polluted habitats may be caused by the absence of their fish hosts, which may be more susceptible to low oxygen than the mussels. The presence of dense mussel beds indicates clean (but not necessarily drinkable) water, partly because the substantial amount of oxygen required over the long term is obviously available and partly because the mussels themselves filter and purify the water. High mussel diversity also indicates high fish diversity, and implies good fishing.

The excellent recent volume edited by Hart and Fuller (1974) gives a more detailed picture of the relationship between molluscs and water pollution.

The Freshwater Molluscs

The following section contains illustrations, range maps, descriptions, and comments about the distribution, ecology, and biology of all the species and subspecies of freshwater molluscs known to inhabit Canada. A few that have not been recorded in Canada but are likely to migrate here are also included.

Many species, especially freshwater mussels used in the manufacture of pearl buttons, have long-standing common names. Such names are given here. For species that do not have them, I have proposed common names. In doing so, I tried not simply to translate the scientific names but to suggest names that give useful information about the morphology, ecology or distribution of the species.

The structural terms used in the key and in the text are defined on the endpapers and in the glossary.

The key that follows is meant to facilitate identification of unknown molluscs by leading the collector directly to the families they belong to. As familiarity with freshwater molluscs grows, the family is identified more readily, but, to begin with, use of the key will be helpful. The next step in identification should be to compare the illustrations of the species in that family and examine the range maps. Finally, by reading the texts on description and habitat, the collector should be able to pinpoint the species the specimen belongs to.

The specimens illustrated in the black-and-white plates that face each of the species descriptions are from the collection of the National Museum of Natural Sciences, Ottawa, and were collected over many years in Canada and the northern United States. The caption under each plate indicates either the specimens' actual measurements or the percentage enlargements. For plates 1 to 90, the captions give the largest dimension of all the specimens illustrated, but in plates 91 through 144 the measurements are those of only the largest of the two pairs depicted. The pea clams illustrated in plates 160 to 179 were photographed through a scanning electron microscope, and only the magnifications are given.

The colour plates are reproductions of watercolours painted from actual specimens especially for this book. The specimens appear in taxonomic order on the plates, and are identified by their scientific name and by the number allotted to that species or subspecies in the text. Information on where the specimens were obtained can be found in the corresponding species descriptions, under the captions for the black-and-white plates. The specimens are reproduced 2/3 of actual size, except for No. 136, which is life size.

It is helpful to remember that in all likelihood the first species collected will be those most abundant in the region. For example, among the freshwater mussels or clams of eastern Canada *Elliptio complanata* is dominant, in the prairies *Lampsilis radiata siliquoidea* is the most common, and in British Columbia *Anodonta kennerlyi* and *Margaritifera falcata* are the most abundant. Among the Lymnaeidae, the first species found is usually *Stagnicola elodes*, and among the pill clams (*Pisidium*) probably *P. casertanum*.

Also important to remember is that living land snails are sometimes found in shallow water, and that empty shells of terrestrial snails commonly occur in beach drift. If a Canadian "freshwater" snail does not appear in this book, it is probably a land snail. These can be identified by using Pilsbry (1939-48) or Burch (1962). In some parts of Canada, for example in the Ottawa region and in the vicinity of Hudson Bay, Pleistocene fossil marine shells also occur in beach drift. In a few areas, such as the St. Lawrence River estuary, some freshwater-tolerant marine molluscs also live in close association with true freshwater species. Marine molluscs can be identified by using Abbott (1974) or Bousfield (1960).

The host fishes for freshwater mussels, where known, are recorded under their standardized common names following the list published by the American Fisheries Society (Bailey et al. 1970).

Key to the Families of Canadian Freshwater Molluscs

1. Shell double, i.e. composed of two halves or valves — Pelecypoda (clams or mussels) **2**

   Shell single, i.e. spiral or cap-shaped — Gastropoda (snails) **4**

2. Shell small, 25 mm long or less (except up to 50 mm in *Corbicula*), with small to medium-sized pseudocardinal hinge teeth and with lateral hinge teeth both in front of the pseudocardinal teeth and behind them — **3**

   Shell larger, more than 25 mm long in most specimens, and with hinge teeth not as above — Superfamily UNIONACEA* (3 families) (p. 247)

3. Shell less than 25 mm long; lateral hinge teeth not serrated — Family SPHAERIIDAE (p. 361)

   Shell up to 50 mm long; lateral hinge teeth finely serrated — Family CORBICULIDAE (p. 357)

4. Shell small, cap-shaped, without hinge teeth and with a clearly defined apex that is not near the margin — (freshwater limpets) **5**

   Shell spiral and not as above — **7**

5. Shell more than 8 mm long; Columbia River system only — Family LANCIDAE (p. 93)

   Shell 8 mm long or less — **6**

6. Apex sharply pointed, acute and located posteriorly and to the left; rare — Family ACROLOXIDAE (p. 89)

   Apex blunt and located posteriorly and in the midline or to the right; common — Family ANCYLIDAE (p. 219)

7. Living snail with an operculum — **8**

   Living snail without an operculum — **13**

8. Shell medium-sized to large, i.e. more than 12 mm high, and with a horny operculum — **9**

   Shell smaller than 12 mm or with a calcareous operculum — **10**

---

*The families of freshwater mussels or clams (superfamily Unionacea) are not separable on shell characters alone; therefore, Unionacea is keyed out here as a unit.

9	Width more than half the height	Family VIVIPARIDAE (p. 33)
	Width less than half the height	Family PLEUROCERIDAE (p. 81)
10	Shell more than 9 mm high and with a calcareous operculum; eastern Canada	Family BITHYNIIDAE (p. 77)
	Shell less than 9 mm high or with a horny operculum	11
11	Shell wider than high, with or without strong spiral ridges, and with a nearly circular aperture; operculum multispiral	Family VALVATIDAE (p. 41)
	Shell higher than wide, without strong spiral ridges, and with an ovate aperture; operculum paucispiral	12
12	Shell small, up to 5.5 mm long, slender (width divided by height about 0.55–0.60), with up to 7 whorls, an ovate aperture, and a distinct lip that surrounds the aperture; amphibious; southeastern Canada only	Family TRUNCATELLIDAE (p. 73)
	Shell relatively wider or with fewer than 7 whorls or without a distinct lip that surrounds the aperture; wholly aquatic; widespread	Family HYDROBIIDAE (p. 55)
13	Shell disc-shaped or spire projecting only a little above body whorl; shell width greater than height	Family PLANORBIDAE (p. 175)
	Spire projecting much above body whorl; shell width less than height	14
14	Shell dextral or right-handed	Family LYMNAEIDAE (p. 97)
	Shell sinistral or left-handed	Family PHYSIDAE (p. 151)

Class Gastropoda (Snails)

Subclass Prosobranchia (Gill-breathing Snails)
Order Mesogastropoda

I Superfamily Viviparacea

FAMILY VIVIPARIDAE (Mystery Snails)

Shells large, dextral, with an elevated spire, non-umbilicate or with a small umbilicus, rather inflated, and mostly smooth. Operculum large, horny and with very few turns (paucispiral). Tentacles long and slender, but in the male the right one is shorter and forms a sheath for the penis. Radula normally with 7 simple or serrated teeth in each row. Central tooth large and broad. Some species are hermaphroditic and others have separate sexes. All are ovoviviparous, that is the eggs mature within the parent, and young are held there for a time before they are released. The family is worldwide.

1
Campeloma decisum (Say, 1816)
Brown Mystery Snail

DESCRIPTION

Shell up to about 42 mm high, moderately high-spired (W/H 0.56-0.69), quite thick and heavy, and with up to 7 convex whorls. Early whorls corroded in all specimens except those from lime-rich water. Spire whorls rounded, narrowly shouldered, and separated by incised sutures. Spire angle about 45° to 50°. Aperture ear-shaped, about half as high as the shell, and white to bluish white within. Umbilicus absent. Periostracum dull to glossy, yellowish or greenish brown, and with dark-brown collabral growth rests. Sculpture consists of fine spiral and collabral lines. Operculum brown, rather thin but tough, concave ear-shaped, with a nucleus located near the inner margin and numerous concentric growth lines.

This large, common species is unlike any other. *Campeloma integrum* (Say, 1821), which lives in lime-rich, alkaline water, appears to be the same species. It is distinguished by having uncorroded early whorls. (This condition would be expected of snails from hard-water habitats.)

DISTRIBUTION

Occurs from Nova Scotia to Manitoba, and south in the Atlantic coastal plain to the southern Atlantic states. Its precise southern limits are unknown.

ECOLOGY

Found in lakes, canals, and slow-flowing rivers and creeks on muddy or mud and sand bottoms. Burrows just below the surface. Often abundant in enriched or eutrophic habitats. The species is parthenogenetic, and males are absent at least from northern populations. The young are held within the uterus until the 3-whorl stage is reached. Sinistral unborn young are relatively common but very few ever reach adulthood. The radula formula is ordinarily 2-1-1-1-2.

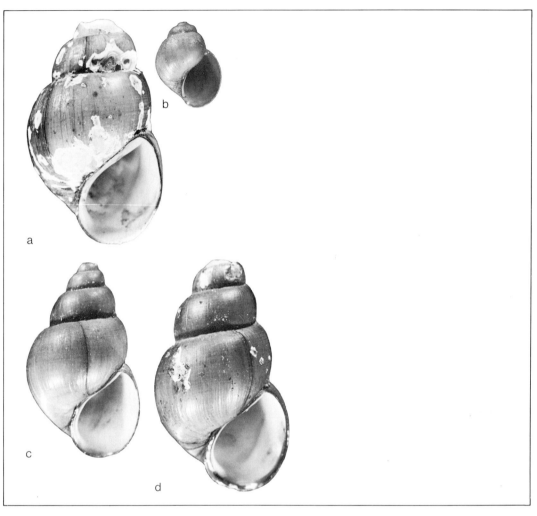

1
Campeloma decisum
a: Chukuni R. near Red Lake, Ont.; 30.6 mm.
b: Ottawa R. near Ottawa, Ont.; 5.8 mm (juvenile).
c,d: Meach L. near Hull, Que.; *c* 26.0 mm, *d* 31.1 mm.

2
Viviparus georgianus
(Lea, 1834)
Banded Mystery Snail

DESCRIPTION

Shell up to about 35 mm high, subglobose (W/H 0.77-0.83), and thin but strong. Nuclear whorl rather large, blunt, and corroded. Whorls 4 or 5, strongly convex, shouldered, and separated by deep sutures. Spire angle about 50° to 65°. Aperture ovate ear-shaped, about half as high as the shell, and with the inner lip partly or fully reflected over the umbilicus and partly or wholly obscuring it. Periostracum pale yellowish to greenish brown, of medium thickness, closely adherent, shining in many specimens, and bearing 3 or 4 prominent dark-reddish spiral bands. Sculpture consisting of coarse and fine growth lines. Operculum thin, corneous, with a subcentral nucleus and concentric growth lines.

Easily identified by its large, globose form and prominent, spiral, dark-reddish bands. The species has often been cited as *V. contectoides* (Binney, 1865).

DISTRIBUTION

Discontinuously distributed (introduced) in parts of the Great Lakes-St. Lawrence River system in Canada (Grand River, Ottawa River system, Richelieu River, and lower St. Lawrence River) and the United States south to Florida and Arkansas. Some northern populations may represent introductions of the European species *V. viviparus* (Linnaeus, 1758), which is virtually indistinguishable from northern populations of the North American *V. georgianus*.

ECOLOGY

Lives in lakes and slow-moving rivers, usually on muddy bottoms and frequently among vegetation. In some favourable localities it is very abundant. The young are held within the uterus until the 3-whorl stage is reached.

2
Viviparus georgianus
a: L. Opinicon, Rideau Lakes, Ont.; 25.5 mm.
b: Bobs L., Rideau Lakes, Ont.; 35.0 mm.

3
Cipangopaludina chinensis (Gray, 1834)
Oriental Mystery Snail

DESCRIPTION

Shell up to about 63 mm high, inflated (W/H 0.74–0.82), globose, and of moderate thickness. Nuclear whorl small, flattened, dark brown, and corroded in many specimens. Whorls about 7, flatly convex, somewhat shouldered, and with prominent sutures. Spire with convex sides and spire angle about 60° to 70°. Aperture ovate ear-shaped, broadest below, about half as high as the shell and bluish white within. Inner lip reflected and concealing most of the umbilicus. Periostracum thick, greenish to reddish brown, smooth and shiny (on most specimens) on upper whorls but roughened by heavy growth rests on the body whorl. Sculpture consists of collabral striae and growth rests and, on many specimens, shallow malleations. Young specimens are strongly carinate at the periphery of the body whorl.

This is the largest freshwater snail in Canada and is best recognized by its relatively gigantic size and brownish periostracum, which lacks prominent spiral bands. *Viviparus japonicus* (von Martens, 1860) and *V. malleatus* (Reeve, 1863) are synonyms. Compare with *V. georgianus* (Lea).

DISTRIBUTION

Introduced from Asia to a few localities in Canada: a pond in lower Sackville, Nova Scotia, a marsh on Île Perrot near Montreal, the Rideau River at Ottawa, a reservoir at St. Thomas, Ontario, and in British Columbia at Victoria, Harrison Mills, and a lake on Saltsprings Island. It probably occurs elsewhere as well. Many colonies are known in the United States.

ECOLOGY

Typically a species of muddy ponds and lakes, backwaters, sloughs, canals, and slow-moving rivers. Like other species in the family, the eggs hatch within the uterus and young are retained there for a considerable period.

3
Cipangopaludina chinensis
a: L. Erie, Sandusky, Ohio; 56.2 mm (shell cleaned).
b: Slough near Klamath Falls, Oreg.; 47.6 mm.
c: Pond, Lower Sackville, N.S.; 6.6 mm (juvenile).

II   Superfamily Valvatacea

FAMILY VALVATIDAE (Valve Snails)

Shells small, dextral, with a moderately elevated spire or flatly-coiled, umbilicate, whorls rounded and smooth, with collabral sculpture (i.e. with lines parallel to former positions of the mouth of the shell during shell growth), or with spiral cords. Operculum horny and with many turns (multispiral). Tentacles long and slender. Gill external and feather-like. Radula with 7 teeth in each row (formula 3-1-3) and teeth with many cusps. Hermaphroditic. Egg capsules are spherical and contain from 1 to 60 eggs. Developing embryos are green. The family occurs in North America, Europe, Asia, and Africa.

4
Valvata perdepressa
Walker, 1906
Flat Valve-Snail

DESCRIPTION
Shell up to about 3 mm high and 6 mm wide but variable (W/H 1.8-2.3), planorboid to flattened above and decurrent below, relatively solid, with about 3-1/2 whorls, and a shining and smooth or finely striated surface. Spire flat and level with, or only slightly above, the body whorl. Nuclear whorl finely punctate and, like the following 1 to 2 whorls, reddish, reddish brown, or brown (in most specimens). Sutures impressed. Whorls flattened above and rounded below. Aperture nearly round except slightly flattened above and in contact with the previous whorl over a narrow area. Umbilicus broad and clearly exhibiting the inner 2/3 of all whorls.

More flattened than any other *Valvata* and has a broader umbilicus. If the reddish early whorls are visible the species may be easily recognized. Compare with *V. sincera helicoidea*.

DISTRIBUTION
Lake Michigan to Lake Ontario and the Lake Ontario drainage area in northern New York State.

ECOLOGY
Has been found only in beach drift on the shores of large and medium-sized lakes. Its detailed ecology and life history are unknown.

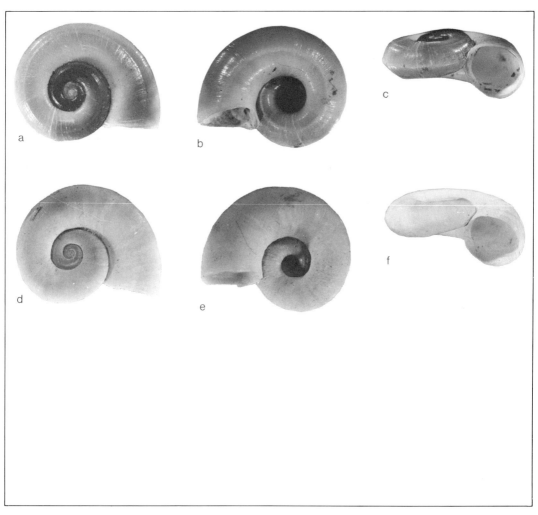

4
Valvata perdepressa
a,b,c: Rondeau Harbour, L. Erie, Ont.; 5.6 mm.
d,e,f: Another specimen, same locality; 5.6 mm.

5
Valvata piscinalis
(Müller, 1774)
European Valve Snail

DESCRIPTION
Shell up to about 5 mm high and 5 mm wide but variable (W/H 0.80-1.20), rather solid, with up to 5 convex whorls, with fine collabral threads and, in some specimens, with partly obscure spiral striae. Spire conical, with an obtuse apex, impressed sutures, and rounded whorls. Aperture circular, except where it is appressed to the penultimate whorl, and in height measuring about 1/2 to 2/3 the height of the shell. Inner lip slightly reflected over a deep, narrow umbilicus. Periostracum thin, yellowish brown to greenish, and shining. Operculum circular and with numerous whorls.

Distinguished from *V. sincera sincera* by its larger size, relatively higher spire, and narrower umbilicus.

DISTRIBUTION
A Eurasian species introduced into North America. It was first noticed in Canada at Toronto in 1913, and has since spread throughout Lake Ontario, and from there into Lake Erie and the upper St. Lawrence River and some of its tributaries.

ECOLOGY
Occurs in lakes and slow-moving rivers in North America as in Europe. Egg masses are deposited on various species of aquatic plants and from 4 to 60 eggs per mass have been observed. The young emerge in 15 to 30 days.

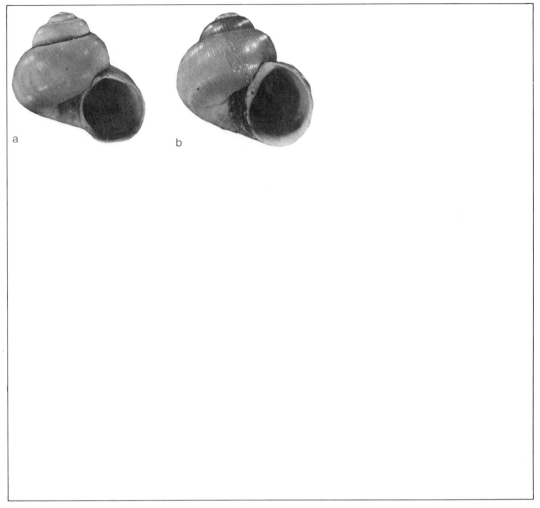

5
Valvata piscinalis
a: Bay of Quinte, L. Ontario, Ont.; 5.3 mm.
b: L. Ontario near Hamilton, Ont.; 5.2 mm.

6
Valvata sincera sincera
Say, 1824
Ribbed Valve-Snail

DESCRIPTION

Shell up to about 3.2 mm high and 5.0 mm wide (W/H 1.3-1.9), moderately solid, with 4 rounded whorls, and widely spaced collabral threads. Spire moderately elevated. Nuclear whorl planorboid and finely striate. The calcareous shell surface of later whorls is covered with rather widely spaced collabral threads (8 or less per mm). Aperture round and in contact with the previous whorl but not flattened by it. Umbilicus round, medium-sized, and deep. Periostracum pale brown to brown; in many specimens, when fresh, it bears fine blade-like collabral lamellae surmounting the collabral threads on the underlying shell. Sutures well-marked and prominent. Operculum corneous, round, and multispiral, with about 6 whorls.

Similar to the more northern *V. s. helicoidea*; however, that subspecies has finer striae (typically 14 or more per mm), is larger, and has a lower spire.

DISTRIBUTION

Newfoundland to British Columbia and the Yukon Territory, and Maine to Minnesota. Intergrades with *V. s. helicoidea* occur within the broad zone of overlap. This zone lies on both sides of the common boundaries shown on Maps 6 and 7. All living non-carinate *Valvata* in British Columbia and the Yukon Territory appear to belong to this species despite literature records of *V. mergella* Westerlund and *V. virens* Tryon.

ECOLOGY

Occurs principally in lakes, often at considerable depths, and usually on mud among submersed aquatic vegetation. Occasionally found in slow-moving rivers and in muskeg pools. The radula is similar to that of *V. tricarinata*.

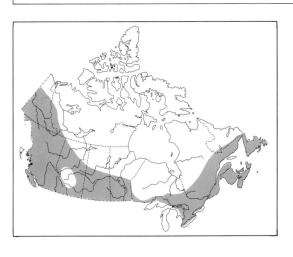

6
Valvata sincera sincera
a: L. Nichicun, Que.; 4.6 mm.
b: Stream near Finland, Ont.; 4.4 mm.

7
Valvata sincera helicoidea Dall, 1905
Northern Valve Snail

DESCRIPTION
Shell similar to that of *Valvata sincera sincera* except that it is somewhat larger, up to 7 mm wide rather than 5 mm, and the collabral sculpturing is finer, more crowded (14 or more riblets per mm), and typically like the winding of thread on a spool. Blade-like periostracal lamellae do not occur in this subspecies. In some specimens the surface is smooth. Many specimens are also more depressed and have relatively wider umbilici than in typical *V. sincera*.

DISTRIBUTION
Labrador to British Columbia, and Alaska north to a line running from Ungava Bay to southern Victoria Island in the arctic archipelago. Principally an arctic and subarctic subspecies.

ECOLOGY
Occurs in lakes, ponds, slow-moving rivers and streams, and in muskeg pools, usually among aquatic vegetation and on a variety of substrates. Lives at various depths down to 15 m, and is often found in the stomachs of whitefish (*Coregonus*). The radula is similar to that of other *Valvata* species.

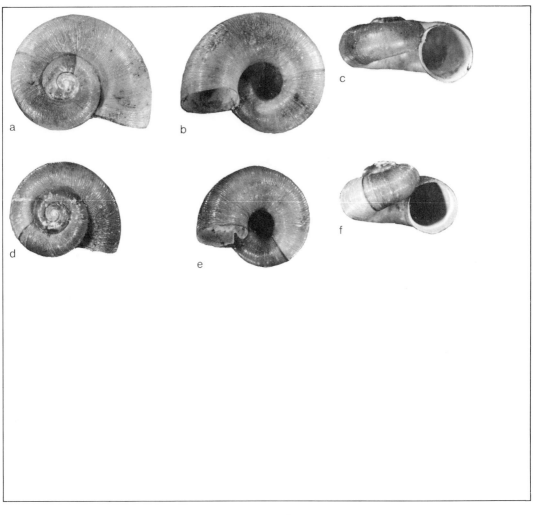

7
Valvata sincera helicoidea
a,b,c: Ennadai L., N.W.T.; *a* 5.1 mm, *b* and *c* 5.4 mm.
d,e,f: Owl L., Man. (56°24′N, 94°55′W); 4.3 mm.

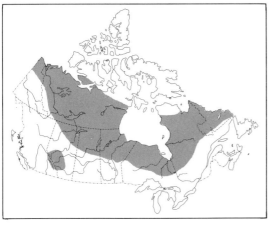

8
Valvata sincera ontariensis
Baker, 1931
Loosely Coiled Valve Snail

DESCRIPTION
Shell similar to *Valvata sincera sincera* except that the body whorl, and in some specimens also the lower half of the penultimate whorl, are not peripherally attached to the preceding whorl. The uncoiled portion is deflected downward and outward to the extent that, near the aperture, the space between the body whorl and the penultimate whorl is approximately twice the body-whorl diameter. The diameter of the aperture is also less than in *V. sincera sincera*. The collabral threads are prominently raised and widely separated.

The partially uncoiled shell of this subspecies is entirely distinctive.

DISTRIBUTION
Known only from Lake Superior (Shakespeare Island Lake) and the region north of Lake Superior drained by the headwaters of the Attawapiskat, Albany and Severn River systems.

ECOLOGY
Occurs in large lakes and slow-moving portions of large rivers at depths of 1 to 10 m, on muddy bottoms, and among aquatic vegetation. Nothing is known about its reproduction. The animal moves forward with the shell stationary, and then hitches the shell suddenly forward.

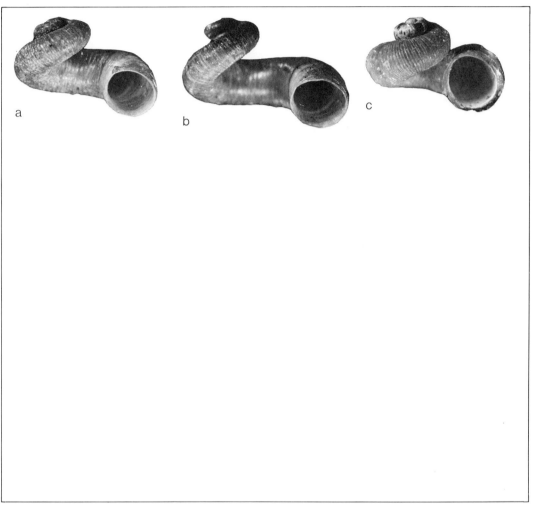

8
Valvata sincera ontariensis
a,b: Klotz L. near Longlac, Ont.; *a* 3.5 mm, *b* 4.1 mm.
c: Lake near Schreiber, Ont.; 3.4 mm.

9
Valvata tricarinata (Say, 1817)
Three-keeled Valve Snail

DESCRIPTION
Shell up to about 5 mm high and 6 mm wide (W/H 1.1-1.6), solid, with 4 whorls, dextral, variable in sculpturing but typically with 3 prominent spiral cords on the body whorl, and with a round aperture. Nuclear whorl planorboid and microscopically striate. Later whorls bearing 3 prominent carinae (in most specimens), 1 at the shoulder, 1 at the periphery and 1 bounding the umbilicus, but on the spire the 2 lower carinae are obscured by the following whorl. On some specimens 1 or more of the cords are reduced to spiral angulations. Whorls flattened between carinae. Sutures impressed. Aperture round and lip continuous. Umbilicus round, funnel-shaped, and deep. Periostracum brown to green. Operculum horny, round, and multispiral with about 10 turns.

This species is very distinctive, and even where the spiral cords are missing the presence of spiral angulations is characteristic.

DISTRIBUTION
Found from New Brunswick to eastern British Columbia, and in the Northwest Territories south of the tree line. In the United States it occurs south to Virginia, Iowa and Nebraska.

ECOLOGY
Occurs among vegetation and only in perennial-water habitats, namely lakes, rivers, streams and muskeg pools. It is rare in ponds. Egg masses are deposited on aquatic vegetation, detached leaves of deciduous trees, and available smooth surfaces. From 4 to 18 eggs have been observed in each mass, and juvenile emergence takes place in 12 to 15 days. The animal is white or pinkish white, and has one feather-like external gill on the left and one that is stalk-like on the right. The radula has 3 lateral teeth on each side of the central tooth, and each tooth bears numerous tiny cusps.

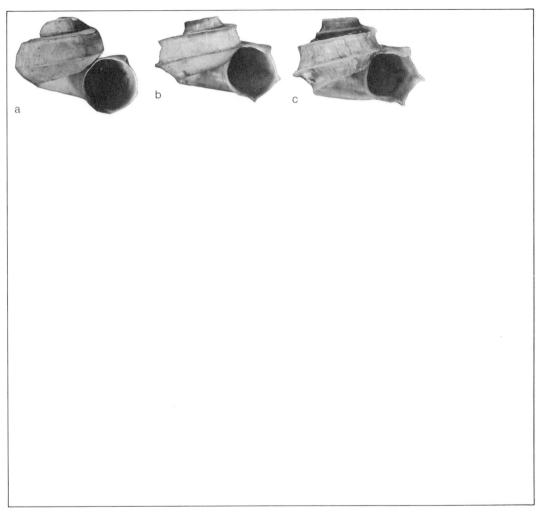

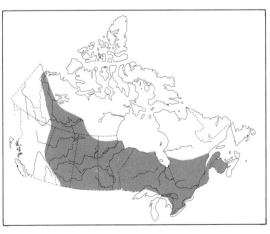

9
Valvata tricarinata
a: Seine R. near Winnipeg, Man.; 4.8 mm.
b,c: Halkett L., Prince Albert National Park, Sask.;
   b 4.8 mm, *c* 4.9 mm.

III Superfamily Rissoacea

FAMILY HYDROBIIDAE (Spire Snails)

Shells small, dextral, with an elevated spire, umbilicate or non-umbilicate, and smooth or sculptured. Operculum horny and paucispiral. Tentacles long and cylindrical, gills internal, and penis located a distance behind the right tentacle. Radula with 7 teeth in each row (formula 2-1-1-1-2) and teeth with many cusps. Dioecious, that is some individuals are male and others are female. Eggs round or oval and attached singly to stones or vegetation. The family is worldwide and has marine, brackish-water, and freshwater species.

10
Cincinnatia cincinnatiensis (Anthony, 1840)
Campeloma Spire Snail

DESCRIPTION
Shell up to about 6 mm high, relatively broad (W/H 0.65-0.82), globose-conical, with up to 6 whorls, and relatively strong. Spire rather broad but acute and pointed. Nuclear whorl small, initially planorboid but then decurrent and projecting above the following whorl. Later whorls roundly convex, slightly shouldered above, and separated by deep sutures. Aperture roundly ear-shaped, narrowed above, and with a continuous lip. Umbilicus prominent, deep, and only a little obscured by the inner lip. Periostracum brown and thin. Sculpture consists of growth rests and fine collabral lines. Operculum thin and paucispiral.

This rare species may be recognized by its pointed apex, open umbilicus, and rather large size; a shell with 4 whorls will be more than 4.5 mm high. Resembles an uncorroded *Campeloma decisum* but in miniature. Compare with *Bithynia tentaculata*.

DISTRIBUTION
Southern Ontario and Pennsylvania west to southern Saskatchewan, Utah and Texas.

ECOLOGY
Occurs in lakes and rivers and on muddy or sandy bottoms. Lives in deeper water than most other hydrobiids. The male reproductive anatomy is distinctive: the verge is bifid with a short penis, much larger secondary lobe, and no accessory duct.

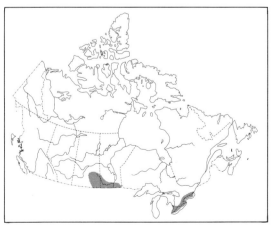

10
Cincinnatia cincinnatiensis
a,b: Echo L., Sask.; 4.9 mm
c: Traverse L., Brown's Valley, Minn.; 5.9 mm.

11
Probythinella lacustris
(Baker, 1928)
Flat-ended Spire Snail

DESCRIPTION
Shell up to about 5 mm high, moderately wide (W/H 0.63-0.84), subcylindrical, and rather solid. About 5 whorls with the first 2 approximately level with, or sunken below, the third whorl. Succeeding whorls flatly convex, increasing in diameter rather slowly, and forming an almost barrel-shaped spire. Sutures deep. Aperture subovate, about 40% the height of the shell and with a continuous lip. Umbilicus moderately narrow and deep. Periostracum thin and brownish. Sculpture consists of growth rests, fine collabral lines, and faint spiral striae. Operculum thin and paucispiral.

Distinguished by its truncated, barrel-shaped spire. Compare with *Amnicola limosa*.

DISTRIBUTION
Quebec to the western Northwest Territories and Alberta; New York to Iowa, Arkansas and Kentucky.

ECOLOGY
Lives in perennial-water lakes, ponds, and rivers of all sizes. Most frequently found among vegetation and on muddy or sandy bottoms. Tends to occur in deeper water in the southern part of its range. The animal is white, and the marginal radula teeth bear no cusps. The verge is bilobed but not bifid, the penis is larger than the secondary lobe, and there is no accessory duct.

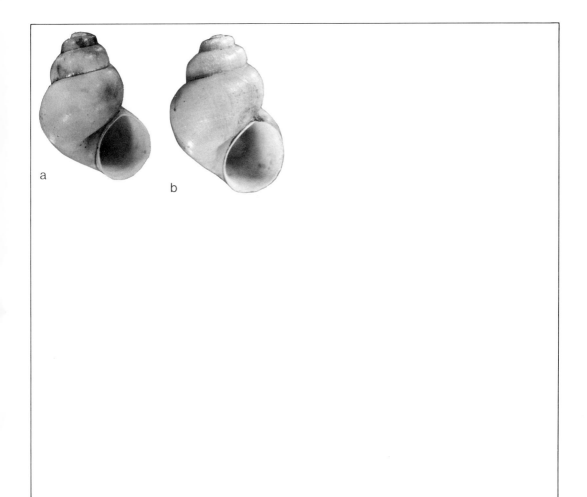

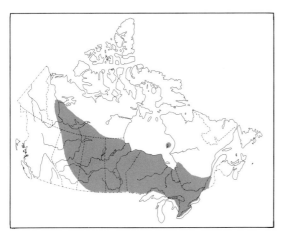

11
Probythinella lacustris
a: L. Erie, Rondeau Park, Ont.; 3.6 mm.
b: L. Athabaska, Alta.; 3.9 mm.

12
Marstonia decepta (Baker, 1928)
Pilsbry's Spire Snail

DESCRIPTION
Shell up to about 4.5 mm high, relatively broad (W/H 0.68-0.84), attenuated, quite thin-shelled, and with elevated apex. Nuclear whorl slightly raised above second whorl, giving the apex a roundly pointed aspect. Whorls number about 4-1/2, roundly convex and, in some specimens, slightly shouldered. Sutures deep. Aperture rather small, ovate, angled above, typically less than half the height of the shell, and with a continuous lip. Umbilicus open but narrow. Periostracum thin and pale brown. Sculpture consists of growth rests and fine collabral threads. Operculum thin, paucispiral, and ovate.

This small species may be distinguished by its elevated apex, narrow umbilicus, and relative size. An *M. decepta* of 4 whorls is 3.5 to 4.5 mm high. *Amnicola walkeri* with 4 whorls is less than 3.5 mm high, while *Cincinnatia cincinnatiensis* with 4 whorls is more than 4.5 mm high. Juvenile specimens are similar to *Pyrgulopsis letsoni* (Walker, 1901), a small hydrobiid sometimes found washed up on Lake Erie beaches (probably fossil). Nevertheless, *P. letsoni* occurs alive in New York and Michigan, and may live in Canada.

DISTRIBUTION
Throughout the Great Lakes-St. Lawrence system; northward in adjacent portions of the Hudson Bay drainage in Ontario; and southward in nearby parts of the Mississippi River system.

ECOLOGY
Occurs with *Amnicola limosa* in permanent lakes, ponds and slow-flowing streams, among vegetation or on rocks. The radula is relatively smaller than in other hydrobiids. The verge is broad distally, not bifid, the penis projects from the right corner, and there is no accessory duct. The eggs are round and lack the lamellar crest seen in eggs of *A. limosa*.

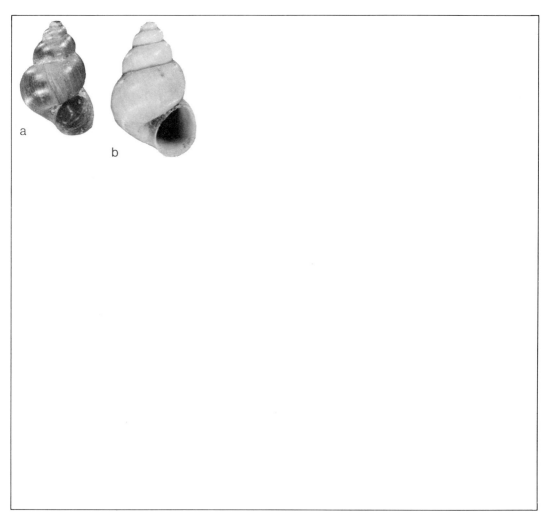

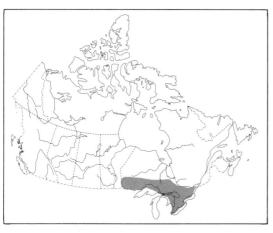

12
Marstonia decepta
a: Wabaskang L., Ont.; 2.8 mm.
b: L. Ontario near Consecon, Ont.; 3.3 mm.

13
Lyogyrus granum (Say, 1822)
Rusty Spire Snail

DESCRIPTION
Shell very small, up to about 2.2 mm high, relatively broad (W/H ca. 0.85-1.00), subglobose, and with about 4 whorls. Spire short. Nuclear whorl flat or descending only slightly, and successive whorls progressively more decurrent; this causes the apex to be bluntly rounded and the spire to bulge on its sides. Whorls convex and sutures deep. Aperture almost circular and barely in contact with the preceding whorl. Umbilicus prominent, deep, and of medium width or narrower. Periostracum covered with a reddish-brown deposit; removal of deposit reveals a brownish to yellowish periostracum and fine collabral and spiral lines. Operculum brown, circular, and multispiral as in *Valvata*.

This tiny snail is well characterized by its bulging spire, deep umbilicus, rusty-brown covering, and multispiral operculum.

DISTRIBUTION
New Brunswick to Virginia in the Atlantic Coastal Plain. Its westward limit is unknown.

ECOLOGY
Sporadic in its occurrence but its colonies may be very dense. Found on the leaves of various species of submersed aquatic plants, or on dead tree leaves on the bottom, in still or slow-moving perennial-water habitats. Nothing has been published regarding its biology.

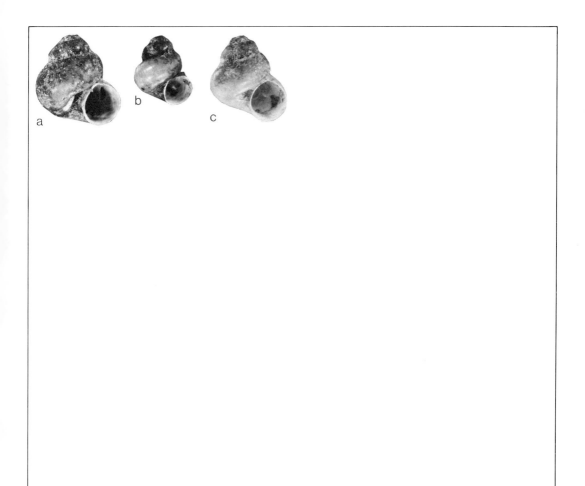

13
Lyogyrus granum
a: Baker Brook near Lincoln, N.B.; 2.2 mm.
b,c: Jemseg R. near Jemseg, N.B.; *b* 1.7 mm, *c* 2.0 mm.

14
Amnicola limosa (Say, 1817)
Ordinary Spire Snail

DESCRIPTION
Shell small, up to about 4.5 mm high, relatively broad (W/H 0.70-0.90), and with about 4-1/2 convex, slightly shouldered whorls. Spire blunt. Nuclear whorl flatly coiled and not projecting above the next whorl. Later whorls rounded, somewhat shouldered, and increasing slowly in size. Sutures deep. Aperture ovate, narrower at the top, and with a thin callus on the parietal wall. Umbilicus deep and of medium width. Periostracum reddish brown, greyish brown, or tan. Sculpture consists of fine, crowded lines of growth. Operculum suboval, thin, pale yellowish-brown, with about 2-1/2 turns, and with spiral and transverse striae.

This small, abundant species may be distinguished by its semiglobose shape and flatly coiled (but not sunken) nuclear whorl. Compare with *A. walkeri* and *Cincinnatia cincinnatiensis*.

DISTRIBUTION
Newfoundland westward, within the tree line, to northern Manitoba and Saskatchewan. In the United States it extends to Florida, Texas and Utah.

ECOLOGY
Occurs in all kinds of unpolluted, perennial-water habitats where aquatic vegetation grows. The radula formula is 2-1-1-1-2, and each tooth is armed with numerous cusps. The verge is bifid, with a pointed penis, blunt secondary lobe, and an accessory duct. The living animal is white or pinkish and quite attractive.

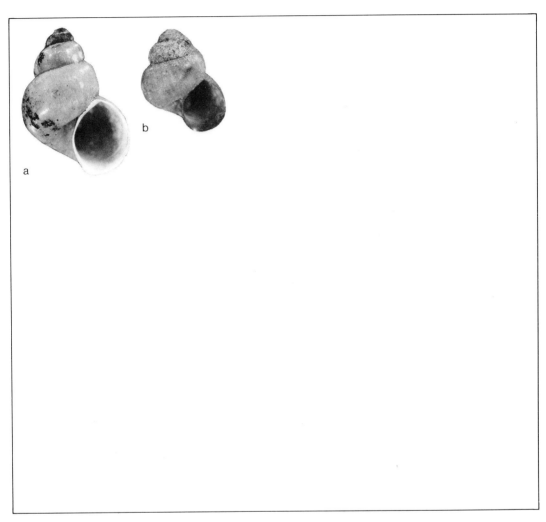

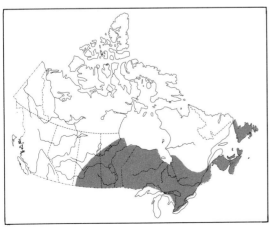

14
Amnicola limosa
a: Black R., Matheson, Ont.; 5.5 mm.
b: L. de Montigny near Val-d'Or, Que.; 4.0 mm.

15
Amnicola walkeri Pilsbry, 1898
Small Spire-Snail

DESCRIPTION
Shell very small, up to about 2.5 mm high, relatively broad (W/H 0.70-0.90) and with a roundly pointed apex. About 4-1/4 convex whorls, with nuclear whorl rounded and projecting slightly above second whorl. The following whorls are sharply rounded and separated by a deep suture. Aperture nearly round and attached to the penultimate whorl at the top and over only a short distance. Umbilicus conspicuous, wide, and deep. Sculpture consisting of numerous fine, crowded collabral lines. Operculum thin, pale, and paucispiral.

Much smaller than *A. limosa* when specimens with the same number of whorls are compared. The apex is also roundly pointed rather than blunt, the whorls are more evenly convex, the aperture is attached to the penultimate whorl over a smaller area, and the umbilicus is relatively wider.

DISTRIBUTION
Throughout the Great Lakes-St. Lawrence drainage area, the upper Mississippi River drainage system, and in the Canadian Interior Basin in the Albany and the upper Nelson River systems.

ECOLOGY
This tiny, uncommon species occurs among dense stands of aquatic plants growing on a muddy bottom in perennial-water lakes and slow-flowing streams. The living animal is heavily pigmented and greyish. The radula is similar to that of *A. limosa* but with a few more cusps. The verge is also similar to that of *A. limosa* except that the lobes are longer.

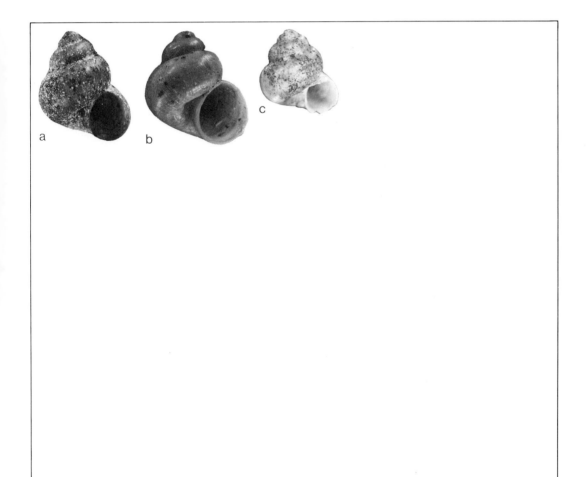

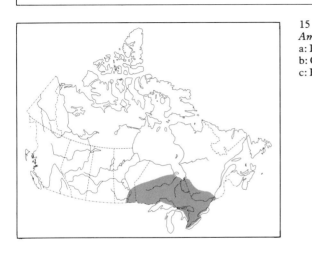

15
Amnicola walkeri
a: Lake near Geraldton, Ont.; 2.8 mm.
b: Cranberry L., Ont.; 2.8 mm.
c: Kimmewin L. near Drayton, Ont.; 2.2 mm.

16
Somatogyrus subglobosus (Say, 1825)
Deep-Water Spire Snail

DESCRIPTION
Shell up to about 9 mm high, relatively broad (W/H 0.80-0.92), solid, and with about 4 whorls. Spire short and body whorl capacious. Nuclear whorl small, spirally striate, and almost flatly coiled. Later whorls convex, decurrent, and expanding rapidly in size. Sutures deep. Aperture broad, ovate, acute above, and more than half the height of the shell. Lip thin. Umbilicus closed or narrowly open, and bordered by the erect inner lip. Periostracum yellowish brown. Sculpture consists of coarse growth rests and fine collabral threads. Operculum ear-shaped, brownish, paucispiral, and fairly thick.

Distinguished by its relatively large and strong shell, short spire, capacious body whorl, and horny operculum. Compare with *Cincinnatia cincinnatiensis* and *Campeloma decisum*.

DISTRIBUTION
Great Lakes-St. Lawrence system and Ohio-Mississippi system throughout.

ECOLOGY
Rare. A rather deep-water species found only in large lakes and large rivers. The radula is relatively large and the central tooth has 3 basal denticles.

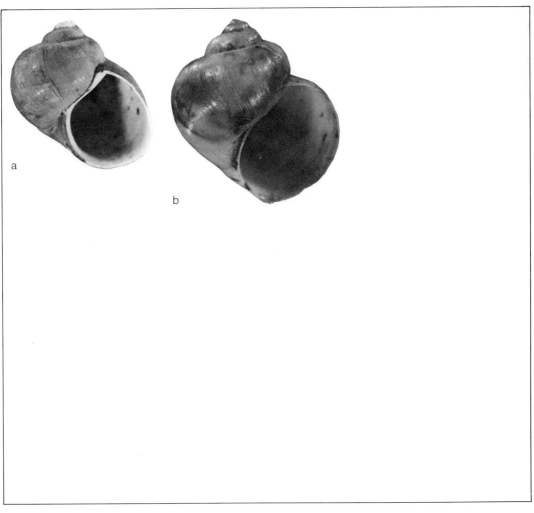

16
Somatogyrus subglobosus
a: L. Saint-Louis, Bellevue Is., Vaudreuil Co., Que.; 5.8 mm.
b: Erie Canal, Mohawk, N.Y.; 7.0 mm.

17
Lithoglyphus virens (Lea, 1838)
Giant Columbia-River Spire Snail

DESCRIPTION

Shell up to about 11 mm high, rotund (W/H 0.68-0.78), with 4 to 5 whorls, thick-shelled, and variable in shape. Nuclear whorl flatly coiled initially and then becoming decurrent. Early whorls corroded in many specimens. Following whorls enlarging rapidly, convex, and separated by deep sutures. Body whorl capacious and constituting most of the shell. Aperture rounded below, acute above, with a thickened peristome that is reflexed over the columellar region completely obscuring the umbilicus or, in a few specimens, exposing a tiny slit. Periostracum chestnut-brown to yellowish brown or olive-green, and with or without darker collabral or spiral bands. Sculpture consisting of coarse growth rests, fine collabral lines, and obscure spiral striations. Operculum thin, pale brown, corneous, paucispiral with about 3 turns, and bearing strong radial wrinkles and fine spiral lines.

Distinguished by its rather large, heavy shell and its spiral, corneous operculum. *Lithoglyphus hindsii* (Baird, 1863) is tentatively considered as a synonym. Further research is necessary to confirm its status, but if *L. hindsii* proves to be distinct then the Canadian species will bear that name. The older name *Lithoglyphus* is used here rather than the more familiar name *Fluminicola*, following Taylor (1966).

DISTRIBUTION

Columbia River system from the Kootenay and Wigwam rivers in southern British Columbia south through Idaho, western Wyoming, northern Utah, Washington, and Oregon. Found also in the Olympic Peninsula of Washington.

ECOLOGY

In the northwest United States this species lives on and under rocks and among vegetation in large and medium-sized lakes, in rivers, and in creeks. Currents at its river localities may be rapid to slow.

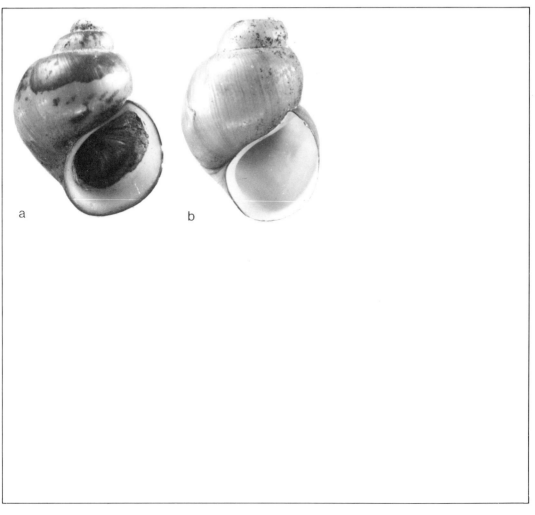

17
Lithoglyphus virens
a: Creek near Olympia, Wash.; 8.1 mm,
b: Siuslaw River, Oreg.; 8.3 mm.

FAMILY TRUNCATELLIDAE
(Looping Snails)

Shells small, dextral, with an elevated spire, more or less cylindrical, umbilicate or non-umbilicate, and smooth or sculptured. Operculum horny and paucispiral. Dioecious. A few species are intermediate hosts for blood flukes, which are parasites of warm-blooded animals, including man. The family is worldwide. Some species live in fresh water, others are amphibious, some live in the litter zone of salt-water beaches, and others are wholly terrestrial.

18
Pomatiopsis lapidaria
(Say, 1817)
River-Bank Looping Snail

DESCRIPTION
Shell small, up to about 5.5 mm high, slender (W/H ca. 0.55–0.60), rather strong, and with about 7 flatly rounded whorls. Nuclear whorl rounded and projecting slightly above the following whorl. Spire high, acute, subtending an angle of about 45°, and with slightly bulging sides. Whorls flatly rounded and separated by incised sutures. Aperture ovate, barely touching the penultimate whorl. Lip thickened and surrounding the aperture. Umbilicus mainly exposed and deep. Periostracum reddish brown to brown. Sculpture consisting of rather fine but prominent and crowded collabral lines.

Similar to *Hydrobia nickliniana* (Lea, 1839) except that in *H. nickliniana* the lip is not thickened, the aperture is larger, and the whorls are more convex. Moreover, *H. nickliniana* is aquatic, not amphibious; it is a doubtful inhabitant of Canada but it may occur in southern Ontario.

DISTRIBUTION
In Canada, this species has been found at only a few places in southern Ontario, namely the Sydenham River at Alvinston, the Thames River near Chatham, and the Ottawa River near Ottawa. In the United States it occupies a broad area from the East Coast to the midwest and south to Texas.

ECOLOGY
Amphibious. Lives on wet ground principally along the edges of streams. Eggs are laid in the soil from spring to late summer. It is capable of serving as the intermediate host of the oriental blood fluke, *Schistosoma japonicum*.

Another species of *Pomatiopsis*, *P. cincinnatiensis* (Lea, 1840), is recorded from southern Ontario by some authors, but I have seen no specimens from Canada. It is relatively wider than *P. lapidaria* and is also amphibious.

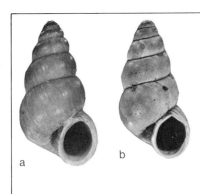

18
Pomatiopsis lapidaria
a: Sydenham R. near Alvinston, Ont.; 5.5 mm.
b: Thames R. near Chatham, Ont.; 5.2 mm.

FAMILY BITHYNIIDAE (Faucet Snails)

Shell small to small-medium, dextral, slender to broad, with an elevated spire, umbilicate or non-umbilicate, and smooth (in most species) or sculptured. Operculum calcareous and paucispiral or with concentric growth lines. Tentacles long, pointed, and tapering. The radula has 7 teeth in each row (fomula 2-1-1-1-2) and each tooth has many cusps. Dioecious. Eggs are laid in groups. The family is worldwide. Previously known as Bulimidae. Bithyniidae and *Bithynia* are now *nomina conservanda*.

19
Bithynia tentaculata
(Linnaeus, 1767)
Faucet Snail

DESCRIPTION
Shell large, up to 13 mm high, moderately inflated (W/H 0.57-0.68), conical, with 5-3/4 whorls, and with a sharply rounded apex. Nuclear whorl smooth, shining, and decurrent. Spire acutely produced. Spire whorls flatly convex and separated by impressed sutures. Aperture less than half the shell height, ovate, but acute above. Lip continuous and somewhat thickened. No umbilicus. Periostracum shining and pale brown. Sculpture consists of growth rests and crowded, fine collabral threads. Operculum white and calcareous.

This common species may be recognized by its large size, calcareous operculum, and lack of an umbilicus. It is often cited as *Bythinia tentaculata* (L) or *Bulimus tentaculatus* (L).

DISTRIBUTION
Introduced and now widespread in the lower Great Lakes-St. Lawrence system, also in the mid-Atlantic United States. Native to Europe.

ECOLOGY
Lives in shallow water in large lakes, large rivers and canals, where it feeds on filamentous algae. Very abundant in favourable situations. The breeding season is in July and August, and egg capsules are deposited principally on the shells of other individuals.

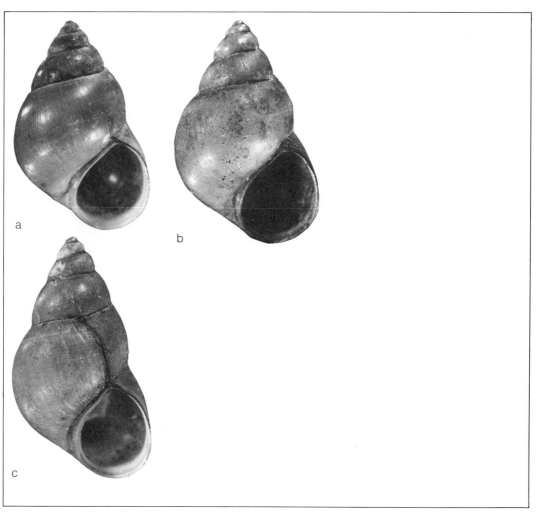

19
Bithynia tentaculata
a,b: Trent R. near Trenton, Ont.; *a* 9.6 mm, *b* 10.1 mm.
c: South Nation R. near Plantagenet, Ont.; 11.0 mm.

IV Superfamily Cerithiacea

FAMILY PLEUROCERIDAE (Horn Snails)

Shell small to medium-sized, dextral, of medium width, high spired, rather heavy-shelled, non-umbilicate, and smooth to highly sculptured. Operculum horny and paucispiral. Tentacles long, tapering, and very narrow; foot short and wide. The radula has 7 teeth in each row (formula 2-1-1-1-2) and each tooth is multicuspid. The animals are dioecious but males lack a penis. Eggs are laid singly or in small groups. Predominantly a North American family, but with a few representatives in eastern Asia.

20
Pleurocera acuta
Rafinesque, 1831
Flat-sided Horn Snail

DESCRIPTION
Shell up to about 37 mm high, narrow (W/H ca. 0.35-0.42), attenuated but variable, and rather thick and heavy. About 14 flat-sided whorls (earliest whorls ordinarily corroded away). Early whorls with 2 spiral carinae but later whorls smooth and forming an even, flat-sided cone. Body whorl angular at the periphery and with or without carinae. Aperture small, about 25% to 30% the height of the shell, with a sigmoid outer lip and a broad canal at the base. No umbilicus. Periostracum blackish, brownish, or yellowish brown. Sculpture, in addition to that mentioned, consists of sigmoid collabral lines. Operculum reddish brown, corneous, and paucispiral with 3 whorls, an acentric, sunken nucleus, and well-marked growth lines and fine striae.

Larger than *Goniobasis livescens*, with flat-sided rather than flatly curved whorls, no callus on the parietal wall (in *G. livescens* a callus is present), a twisted instead of smooth columella, and a body whorl that is angular at the periphery rather than rounded. Also the aperture is relatively smaller, more angulate, and with a broad canal below.

DISTRIBUTION
Great Lakes-St. Lawrence system, upper Ohio-Mississippi drainage, and Erie Canal and contiguous waters in New York State.

ECOLOGY
Found in quiet areas of large streams and in lakes. A burrowing species that prefers mixed sand and mud bottoms. Eggs are deposited in sand-covered masses from April to June.

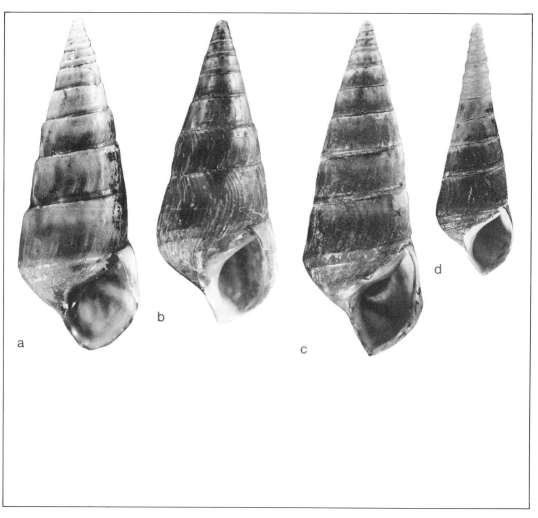

20
Pleurocera acuta
a,b,c: Grand R., Grand Rapids, Mich.; *a* 26.7 mm,
  b 24.6 mm, *c* 27.3 mm.
d: L. Erie near St. Williams, Ont.; 20.2 mm.

21
Goniobasis livescens
(Menke, 1830)
Great Lakes Horn Snail

DESCRIPTION

Shell up to about 25 mm high, rather narrow (W/H ca. 0.38-0.50), more or less attenuated but highly variable in shape, and of medium thickness. Up to about 10 flat, flatly rounded, or rounded whorls (but earliest whorls ordinarily corroded away). Early whorls with a low but prominent spiral carina just above the suture. Later whorls without a carina and, in many specimens, bearing coarse, collabral wrinkles. Suture bordered below by a narrow, pale-coloured band. Aperture sharply rounded below, acute above, about 35% to 40% the height of the shell, and in some specimens brownish within. No umbilicus. Sculpture, in addition to carinae and wrinkles, of fine collabral lines and, in some specimens, also a few low, spiral bands. Colour yellowish, brown or black. Operculum brown, corneous, and paucispiral with 3 whorls, an acentric nucleus and radial striae.

Sometimes confused with *Pleurocera acuta*. See that species.

DISTRIBUTION

Widespread in the Great Lakes-St. Lawrence system. Also in the Erie Canal in New York and in some contiguous water bodies.

ECOLOGY

Occurs in lakes, rivers, streams of all sizes, and springs. Frequently found crawling on stones in a few centimetres of water in clear, rapid streams, but also lives at several metres depth in lakes. Eggs are laid singly or in small groups from April to August.

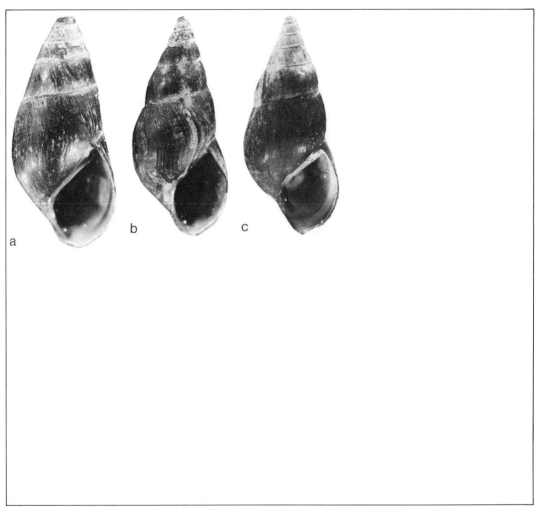

21
Goniobasis livescens
a,b: South Nation R. near Plantagenet, Ont.;
   a 18.7 mm, *b* 17.6 mm.
c: Salmon R., Hastings Co., Ont.; 17.6 mm.

22
Juga plicifera (Lea, 1838)
Graceful Keeled Horn Snail

DESCRIPTION
Shell up to about 35 mm high, narrow (W/H ca. 0.28-0.35), attenuate but variable. About 15 whorls (the early whorls ordinarily corroded away), with about 10 to 12 heavy axial plicae on each whorl (in some specimens obsolete on the latest whorls or confined to the upper halves of those whorls), and with numerous spiral cords. Earliest 3 or 4 whorls very convex and more-or-less smooth. Aperture rounded below, acutely angled above, with a sigmoid outer lip, and with a broad canal below the columella. Periostracum blackish or brown. Sculpture (in addition to above) of heavy sigmoid growth rests and fine collabral threads. Operculum ovate, with about 3 whorls, dark brown except paler above, and with nucleus near centre of lower part.

This is the only species of *Juga* living in the State of Washington and therefore the only one from the northwestern United States likely to be found in Canada.
J. silicula (Gould, 1847) is a synonym.

DISTRIBUTION
Olympic Peninsula of Washington and south in the Columbia and other drainage systems to California. Doubtfully recorded from Vancouver Island. Possibly in southern British Columbia.

ECOLOGY
Occurs in muddy-sand bottoms of small and medium-sized lakes and slow-flowing streams. Like *Pleurocera acuta*, it burrows in the substrate. The animal is pale dusky-grey, with dark transverse stripes on the head and tentacles. Distinguished from *Pleurocera* and *Goniobasis* on the basis of genitalia and egg-mass formation (see Taylor 1966; Clarke 1976).

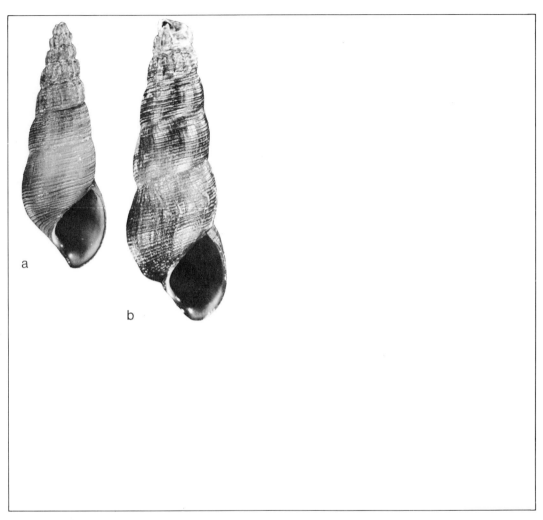

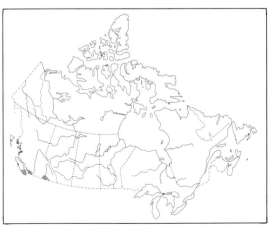

22
Juga plicifera
a,b: Tahkenish L., Douglas Co., Oreg.; *a* 19.7 mm, *b* 24.5 mm.

Subclass Pulmonata
(Lung-breathing Snails)
Order Basommatophora

V Superfamily Acroloxacea

FAMILY ACROLOXIDAE
(Primitive Freshwater Limpets)

Shell small, limpet- or cap-shaped, thin, with an ovate aperture, distinct radial sculpturing, and an acute apex located posteriorly and to the left of the midline. The anatomy is dextral whereas in Lancidae and Ancylidae it is sinistral. The radula has many (about 10 to 15) marginal teeth and about 4 to 7 lateral teeth on each side of the central tooth. The egg capsules are transparent, gelatinous, and contain a small number of eggs. The family is predominantly Eurasian; there is but one North American species.

23
Acroloxus coloradensis
(Henderson, 1939)
Pointed Lake Limpet

DESCRIPTION

Shell up to about 4.6 mm long, 2.9 mm wide and 1.2 mm high, limpet-shaped, thin shelled, elliptical or ovate, and with a prominent, characteristic apex. Apex sharply pointed, pinched, thorn-like, located behind the centre and directed to the left, and radially striate with the striae continuing over the surface of the shell to the margin. Anterior, posterior, and lateral surfaces all flattened, or anterior surface slightly convex and/or posterior surface somewhat concave. Anterior and posterior margins rounded and lateral margins flatly convex and, in some specimens, convergent posteriorly. Periostracum of moderate thickness, brown, and adherent. Surface sculptured with fine radial striae and fine lines of growth.

Can be easily distinguished by its sharp, spine-like apex, which is directed posteriorly and toward the left margin. In other freshwater limpets, the apex is rounded, not pinched, and in the midline or on the right of centre.

DISTRIBUTION

At the present time is known from only a very few localities in the Rocky Mountains, namely Purden Lake near Prince George, British Columbia, Jasper National Park in Alberta, and Montana and Colorado; in eastern Canada only from Matheson and Arkell, Ontario, and the Chibougamau area of Quebec.

ECOLOGY

Rare. Characteristic of rocky, exposed portions of oligotrophic and mesotrophic lakes, where it occurs in shallow water on the undersides of rocks. It has also been found in a eutrophic pond (near Arkell, Ontario) attached to logs. Egg capsules are pale yellow and contain 2 or 3 eggs. The radula formula is about 7-13-1-13-7.

23
Acroloxus coloradensis
a,b: L. Gabrielle, Chibougamau region, Que.; 4.7 mm.
c,d: Pond northeast of Arkell, Halton Co., Ont.;
   4.7 mm.

VI Superfamily Lymnaeacea

FAMILY LANCIDAE
(Limpet-like Lymnaeas)

Shell small-medium, limpet- or cap-shaped, rather thin, with an ovate aperture, concentric and radial striae, and a distinct blunt apex located in the midline and subcentral to, or in front of, centre. A ring-shaped muscle scar is visible in the interior. The anatomy, jaw, and radula are similar to Lymnaeidae, not to Acroloxidae or Ancylidae. The egg capsules are sausage-shaped, transparent, gelatinous, and contain only a few eggs. The family consists of only the genus *Lanx* but there are three subgenera, *Lanx*, *Fisherola*, and *Walkerola*. It is restricted to Pacific coastal drainages from the Columbia to the Sacramento River system.

24
Lanx (Fisherola) nuttalli (Haldeman, 1841)
Greater Columbia-River Limpet

DESCRIPTION

Shell limpet-like, up to about 13 mm long, 10 mm wide, 6 mm high, and with the apex prominent in the midline and placed close to the anterior end. Apex smooth. Anterior and posterior margins more sharply rounded than lateral margins. Anterior slope straight or concave. Posterior slope convex. Interior bluish or purplish in the central portion but whitish around the edge. A subovate muscle scar encircles the inner portion of the shell; this scar is continuous except for a gap on the right side. Periostracum brown to brownish black. External sculpturing consists of concentric growth rests, fine concentric lines, and more or less discernible radial striae.

The relatively large and heavy shell of this species distinguishes it from all other North American limpets. The anterior apex and the discontinuous ring-like muscle scar will distinguish *L. nuttalli* from other species of *Lanx*.

DISTRIBUTION

Columbia River and its tributaries in the northwestern United States. Not yet found living in Canada although recently Dr. Leonard Kalas reported finding a broken shell in the Columbia River at Trail, British Columbia.

ECOLOGY

Occurs in the Columbia River upstream from Richland, Washington, on diatom-covered rocks in the main channel of the river, which is free-flowing in that region. Living specimens are accessible only during periods of low water in the late summer or early fall.

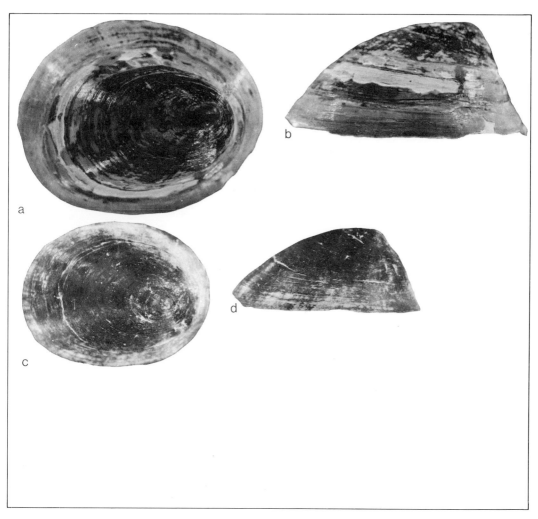

24
Lanx nuttalli
a,b: Columbia R. near McNary Dam, Benton Co., Wash.; 12.2 mm
c,d: Another specimen, same locality; 9.4 mm.

FAMILY LYMNAEIDAE (Pond Snails)

Shell small to large, dextrally coiled, mainly thin-shelled, with an elevated spire, with or without an umbilicus, and without an operculum. Tentacles short, triangular, and flattened. The jaw has 3 plates and the radula has a small central tooth, a few bi- or tricuspid laterals and many bi-, tri- or multicuspid marginals. Monoecious (male and female organs in the same animal) and capable of self- or cross-fertilization. Egg masses are gelatinous, transparent, colourless, sausage-shaped, and contain from a few to many eggs. The family is worldwide.

25
Fossaria decampi (Streng, 1896)
Shouldered Northern Fossaria

DESCRIPTION

Shell up to about 11 mm high, 5 mm wide, with 6 shouldered whorls (most specimens are much smaller) and with aperture about 45% to 60% of the shell height, lymnaeiform, dextral, and of medium thickness. Nuclear whorls satiny, brown, and forming a bluntly rounded apex. In many specimens the brown colour extends to the penultimate whorl and contrasts with the whitish body whorl. Whorls abruptly to roundly shouldered, and separated by deep sutures. Aperture narrow, abruptly arched and rounded above, flattened laterally, and broadly rounded below. Inner lip narrowly reflected and, in its lower part, erect. Umbilical chink prominent and bordered by inner lip. Sculpture consisting of growth rests and numerous fine, closely spaced collabral lines and ridges.

This small species may be recognized by its shouldered whorls, laterally flattened body whorl, elevated and reflected inner lip, and characteristic aperture that is narrowly arched above and broadly rounded below.

DISTRIBUTION

Great Lakes-St. Lawrence drainage north to the Hudson Bay lowlands and the Mackenzie River, and west to British Columbia. Its southern limits have not been determined.

ECOLOGY

A cold-water species occurring only in large lakes in the southern part of its range and in both lakes and rivers in the northern part. It lives among submersed vegetation and on various kinds of bottom. Radula formulae of 20-1-19 to 23-1-22 have been found. The first lateral teeth are tricuspid.

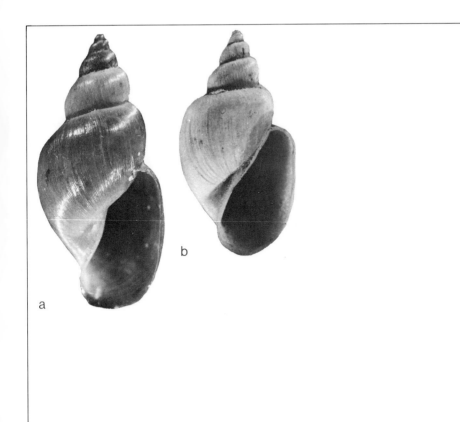

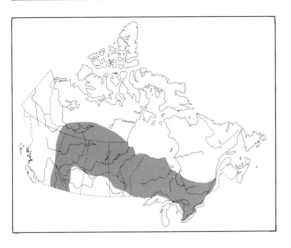

25
Fossaria decampi
a: Halkett L., Prince Albert National Park, Sask.; 11.2 mm.
b: Edith L. near Jasper, Alta.; 9.4 mm.

26
Fossaria exigua (Lea, 1841)
Graceful Fossaria

DESCRIPTION
Shell up to about 9 mm high, 4 mm wide, with 5-1/2 whorls (most specimens are smaller) and with aperture about 45% to 50% of shell height, lymnaeiform, dextral, narrow, attenuated, and thin-shelled. Nuclear whorls satiny and dome-shaped. Spire extended and subtending an angle of about 40°. Spire whorls as high as wide, or higher than wide, and roundly shouldered. Sutures deeply impressed. Body whorl flattened, subcylindrical and about 2/3 the length of the shell. Aperture ovate, slightly flared at base, outer lip thin, and inner lip slightly reflected with the lower part nearly erect. Umbilicus small and narrowly open, or closed by the reflected lip. Periostracum partly or wholly brown. Sculpture consisting of fine collabral lines, irregular growth rests and, in some specimens, poorly defined spiral lines.

This small snail may be identified by its narrow and attenuate form and its tall and roundly shouldered whorls. Compare with *F. modicella* and *F. parva*.

DISTRIBUTION
Great Lakes-St. Lawrence system north to the Hudson Bay lowlands, west to Manitoba and Minnesota, and south in the Ohio-Mississippi system to Alabama.

ECOLOGY
Lives among vegetation in protected parts of lakes and ponds, in backwater areas of rivers, in swamps, and in subarctic muskeg pools. Mud is the usual substrate. Radula formulae of 20-1-21 to 25-1-25 have been recorded.

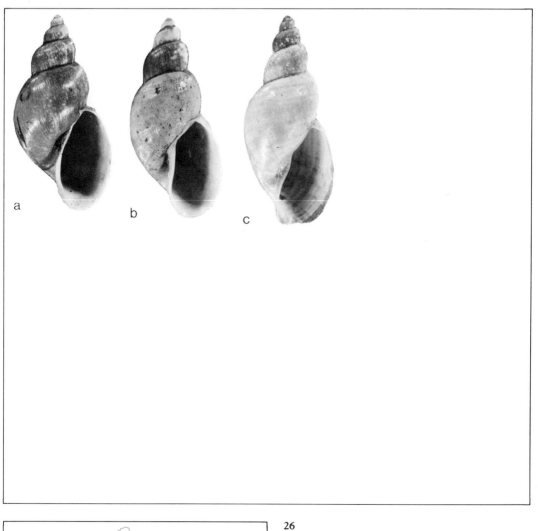

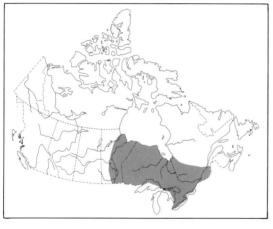

26
Fossaria exigua
a: Lower Red L., Minn.; 7.9 mm.
b: Attawapiskat R. near Attawapiskat, Ont.; 8.1 mm.
c: Winisk R. near Winisk, Ont.; 8.4 mm.

27
Fossaria ferruginea
(Haldeman, 1841)
Fragile Fossaria

DESCRIPTION
Shell up to about 8 mm high, 4 mm wide, with 5 rounded whorls and with aperture about 50% to 55% of shell height, lymnaeiform, dextral and thin-shelled. Spire extended and subtending an angle of about 45°. Sutures impressed. Aperture ear-shaped, outer lip convex and gently rounded above, inner lip narrow and reflected. Umbilicus closed by inner lip or left open only as a small chink. Periostracum yellowish brown or reddish and shining. Sculpture of narrow, clearly defined collabral lines and faint, irregular spiral lines.

Differs from *F. truncatula* in that the whorls are sharply rounded above, the sutures are not deep, the shell is very thin (*truncatula* is strong and solid), and the umbilicus is nearly or completely closed.

DISTRIBUTION
Fraser River drainage (Hatzic Lake) and Columbia River drainage (Duck Lake near Creston) in British Columbia south to central California.

ECOLOGY
The scant information available indicates that it lives among dense vegetation in eutrophic lakes. Other habitats, however, may also be occupied. Nothing is known about its reproduction, radula, or soft anatomy.

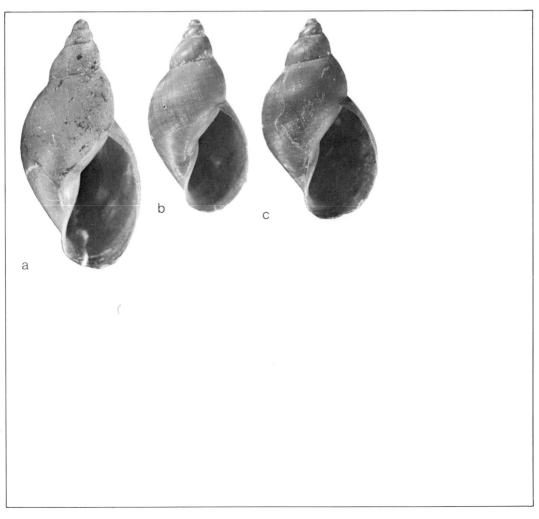

27
Fossaria ferruginea
a: Duck L. near Creston, B.C.; 10.0 mm (spire broken).
b,c: Shuswap L. near Salmon Arm, B.C.; *b* 7.9 mm, *c* 8.2 mm.

28
Fossaria modicella (Say, 1825)
Modest Fossaria

DESCRIPTION
Shell up to about 9.5 mm high, 5 mm wide, with 5-1/2 whorls and with aperture about 45% to 55% of shell height, lymnaeiform, dextral, elongated, and moderately thin-shelled. Nuclear whorls satin-like and bluntly dome-shaped. Spire moderately attenuate and spire angle about 40° to 45°. Whorls flatly rounded, without shoulders, and not as high as wide. Sutures well marked but not deep. Aperture subelliptical. Outer lip thin and convex; inner lip slightly thickened, narrow, somewhat reflected or turned up, and straight or bent just above the umbilicus. Umbilicus small and open. Periostracum light brown or yellowish brown. Sculpture consisting of fine collabral lines and ridges, irregularly spaced growth rests and, in some specimens, numerous fine spiral lines.

Differentiated from *F. exigua* by the absence of whorl shouldering and by the relative height of the whorls (height less than width in *modicella*, equal or greater than width in *exigua*). Differs from *F. parva*, in that the whorls are more flatly rounded, the sutures are not deep, the aperture is elliptical rather than rounded, and the umbilicus is not partly obscured by the inner lip.

DISTRIBUTION
North America south of the tree line, except not recorded from the southeastern United States or Mexico.

ECOLOGY
Occurs in perennial lakes, ponds, and streams, and in vernal pools and ditches. Also occurs on moist sandy or muddy beaches. Vegetation is normally present, and the most commonly observed substrate is mud. The radula formula is reported as 25-1-25 but no doubt some variation exists.

For a discussion of the problematic *Fossaria modicella* morph *rustica* see Clarke (1973) and references cited there.

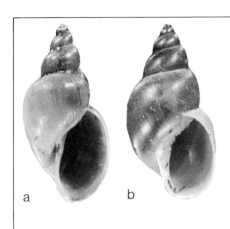

28
Fossaria modicella
a: Kakisa L. near Great Slave L., N.W.T.; 7.2 mm.
b: L. Manitoba, Man.; 7.1 mm.

29
Fossaria parva (Lea, 1841)
Amphibious Fossaria

DESCRIPTION

Shell up to about 8 mm high, 4 mm wide, with 5-1/2 whorls and with aperture about 40% to 50% of shell height, lymnaeiform, dextral, and of moderate thickness. Nuclear whorls satiny, about 1-1/4 in number and forming a rounded apex. Spire angle about 45°. Whorls very convex and rounded with deep sutures. Aperture medium-sized, uniformly ovate or, in many specimens, with inner lip centrally indented and continuous. Outer lip thin and convex; inner lip wide and reflected. Umbilicus clearly defined, deep, and partly covered by the inner lip. Periostracum yellowish brown to brown. Sculpture is of crowded, fine collabral lines, irregular growth rests, and obscure spiral lines.

Distinguished by its roundly convex whorls, incised sutures, rounded aperture, reflected inner lip, and open umbilicus. Its amphibious habits are also distinctive. Compare with *F. modicella, Bakerilymnaea bulimoides* morph *perplexa*, and *B. dalli*.

DISTRIBUTION

Occurs throughout most of North America north of 36° latitude and south of the tree line, except that it is absent from both eastern and western coastal drainage systems. In the Rocky Mountains it extends south to Arizona and New Mexico.

ECOLOGY

Amphibious. Lives on wet mud flats, lake shores and river banks near the water's edge, and in marshes. Also occurs among vegetation submerged in shallow water but is more prone than any other lymnaeid to leave the water. Radula formulae of 16-1-16 to 24-1-24 have been recorded.

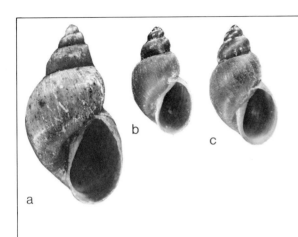

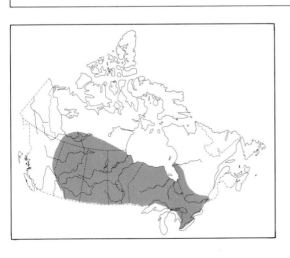

29
Fossaria parva
a: Wainwright Park, Wainwright, Alta.; 7.3 mm.
b,c: Pond near Whitemud Creek, west of Edmonton, Alta.; *b* 4.4 mm, *c* 5.0 mm.

30
Fossaria truncatula
(Müller, 1774)
Liver-Fluke Fossaria

DESCRIPTION
Shell up to about 11 mm high, 6 mm wide, with 5-1/2 rounded whorls (most specimens are smaller) and with aperture about 40% to 50% of shell height, lymnaeiform, dextral, and fairly solid. Nuclear whorls 1-1/4 in number, with the first whorl very small and the next much larger. Spire extended and turreted; spire angle about 45°. Sutures deeply impressed. Aperture ovate; outer lip thin, convex, and abruptly rounded above; inner lip folded back. Umbilicus open and partly covered by the broad reflexed inner lip. Periostracum pale yellowish brown or greyish brown and shining. Sculpture consisting of fine collabral lines and fine, irregularly spaced spiral lines.

The strongly rounded whorls (shouldered in many specimens), solid shell, deep and partly obscured umbilicus, and general appearance are such distinctive features that identification of this species, once seen, can confidently be made. Compare with *F. ferruginea*.

DISTRIBUTION
Columbia River drainage in British Columbia (Osoyoos Lake, stream near Bridesville, Christina Lake, and Duck Lake near Creston), and Yukon River drainage in the Yukon Territory and Alaska. The species is widespread in Europe and in central Asia. It is also reported in Iceland and in scattered localities in Africa.

ECOLOGY
Occurs among vegetation in permanent lakes, ponds, streams, and marshes. Usual substrate is mud. The species is an intermediate host for the sheep-liver fluke, *Fasciola hepatica*. The first lateral radula teeth are tricuspid.

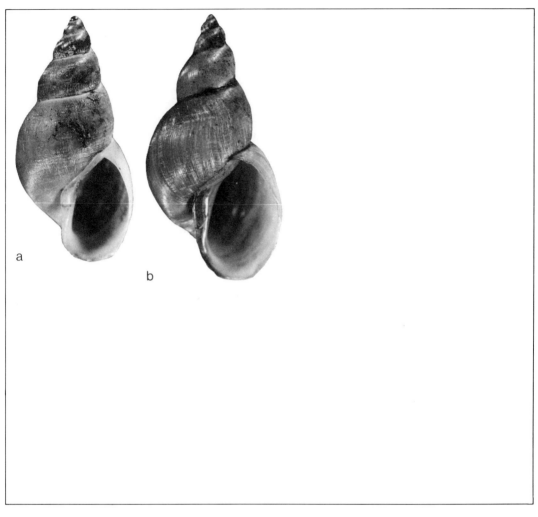

30
Fossaria truncatula
a: Osoyoos L., B.C.; 10.0 mm.
b: Stream near Bridesville, B.C.; 10.7 mm.

31
Bakerilymnaea bulimoides
(Lea, 1841)
Prairie Pond Snail

DESCRIPTION
Shell up to about 11 mm high, 7 mm wide, with 5-1/2 whorls and with aperture length about 40% to 63% of the shell height, lymnaeiform, dextral, and highly variable in shape. The spire may be acutely pointed and pinched (morph *techella*), acute but not pinched (morphs *alberta, perplexa*, and *vancouverensis*), subacute and somewhat bulbous (morph *bulimoides s. str.*), or broad, rounded, and obtuse (morph *cockerelli*). Nuclear whorl smooth, satiny, and brown. Sutures impressed. Body whorl inflated and dominant, especially in morph *cockerelli*. Aperture subovate. Outer lip with a thickened internal varix (reddish-brown in many specimens) behind the edge. Inner lip broad and extending over the umbilicus but not obscuring it. Periostracum brown to greyish brown. Sculpture consisting of fine collabral lines and, in some specimens, also of finer spiral striae. Irregular dark-brown and whitish collabral streaks may also be present.

This highly variable species is best distinguished by its characteristic shell forms (in morphs *techella, bulimoides s. str.* and *cockerelli*, see figures) and by its expanded columella. Morphs *alberta* and *perplexa* (which appear to be synonymous) are similar to *Fossaria parva* but differ in having slightly thicker shells and broader reflected inner lips. Morph *vancouverensis*, a giant form (18 mm) from southern Vancouver Island, is intermediate between morphs *techella* and *bulimoides s. str.* in shell form and may be a distinct subspecies. Examination of the radula may be necessary for positive identification. Species of *Fossaria* have tricuspid first lateral teeth, while those of *Bakerilymnaea* are bicuspid.

DISTRIBUTION
Southern Manitoba to southeastern British Columbia and southern Vancouver Island. In the United States the region west of the vicinity of the Mississippi River.

ECOLOGY
Lives in perennial-water habitats (lakes, ponds, and slow-moving streams) and vernal habitats (roadside ditches, temporary pools). Ordinarily occurs among dense vegetation that is growing on a mud bottom. Radula formulae of 20-1-19 to 23-1-23 have been recorded.

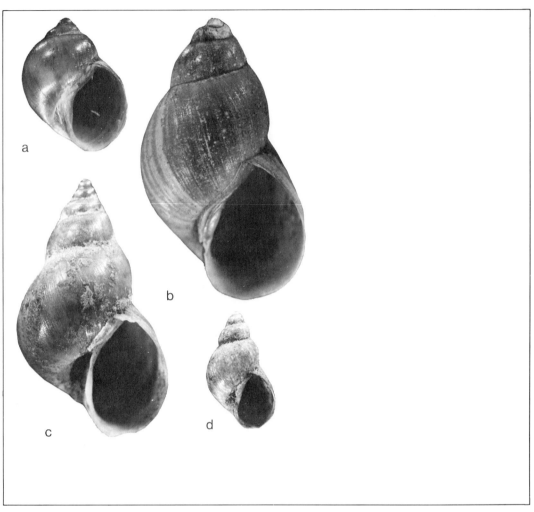

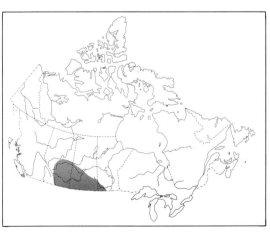

31
Bakerilymnaea bulimoides
a: Morph *cockerelli*: pool near Taber, Alta.; 5.5 mm.
b: Morph *bulimoides s. str.*: pool near Aldersyde, Alta.; 11.4 mm.
c: Morph *techella*: pool near Cayley, Alta.; 10.5 mm.
d: Morph *perplexa*: ditch near Eisenhower Junction, Alta.; 4.7 mm.

32
Bakerilymnaea dalli
(Baker, 1907)
Small Pond-Snail

DESCRIPTION

Shell up to 6 mm high, 3.3 mm wide, with 5 whorls and with aperture about 45% to 55% of shell height, lymnaeiform, and dextral. Whorls convex, roundly shouldered and 5 in number. Nuclear whorl small, flatly rounded, brownish and satiny. Sutures very deep. Spire angle about 50°. Aperture ovate and lip continuous in many specimens. Outer lip evenly rounded. Inner lip flatly reflected over the umbilicus, which is small but clearly defined. Sculpture consisting of fine collabral threads and finer but somewhat obscure spiral striae.

Recognized by the large number of whorls in relation to its very small size, by its incised sutures, and by its roundly shouldered whorls. Compare with *Fossaria parva*.

DISTIBUTION

Southern Ontario and Ohio west and north to the Prairie Provinces and eastern British Columbia, and south to Arizona. Details of its southern distribution have not been worked out.

ECOLOGY

Lakes, ponds, small rivers, and marshes are all suitable habitats for this small species. Vegetation is present and bottom deposits are of diverse types. Does not normally occur out of the water and thereby differs from *Fossaria parva*. Radula formulae of 22-1-21 and 21-1-21 have been recorded. The first lateral teeth are bicuspid.

32
Bakerilymnaea dalli
a: Creek near Edson, Alta.; 5.4 mm.
b: Wigwam R., B.C.; 4.7 mm.

33
Radix auricularia
(Linnaeus, 1758)
European Ear Snail

DESCRIPTION
Shell up to about 30 mm high, 25 mm wide (most specimens are half this size), with 5 whorls, swollen lymnaeiform, dextral, thin, and fragile. Spire short, acute, and pinched in many specimens. Body whorl greatly swollen, enlarged, and constituting more than 90% of the volume of the shell. Aperture ear-shaped, patulous, dilated, and even extending above the spire apex in some specimens. Periostracum thin and pale brown or golden. Sculpture consists of crowded collabral lines, microscopic spiral lines and often irregular malleations or flattened spiral bands.

The bulbous body whorl and small, sharply pointed spire of this large species distinguish it from all others. Compare with *Lymnaea stagnalis sanctaemariae*.

DISTRIBUTION
Native to Eurasia but has been widely introduced in North America, especially, but not exclusively, in the vicinity of major cities. Known to occur in western Canada at Banff, Alberta, and in British Columbia in Kootenay Lake, the Okanagan River system, and North Vancouver.

ECOLOGY
Occurs in lakes, ponds, and slow-moving rivers. Mud is a frequent substrate. The radula formula is approximately 50-1-50, with the first lateral tooth tetracuspid.

33
Radix auricularia
a: Pond, Warm Spring, Park Co., Colo.; 25.0 mm.
b: Lava L., Deschutes Co., Oreg.; 18.8 mm.

34
Radix peregra (Müller, 1774)
Wandering Snail

DESCRIPTION
Shell up to about 7 mm high, 5 mm wide, with 3-1/2 whorls and with aperture height about 75% of shell height, obese lymnaeiform, dextral, and thin-shelled. Whorls flatly rounded. Nuclear whorl rounded, shining and finely punctate. Spire short, broad, and with spire angle exceeding 90°. Body whorl ovate, dominant, and about 90% as long as the shell. Aperture ovate, wide and long, and with thin, broadly curved outer lip and flattened inner lip that bears a thin callus. Umbilicus closed by the reflected inner lip. Periostracum brownish. Surface shining and sculptured by distinct, fine, collabral lines and ridges and obscure spiral lines.

Resembles *Radix auricularia* but North American specimens are much smaller and their apertures are less expanded. In addition, the first lateral radular teeth are tricuspid, not tetracuspid as in *R. auricularia*. Compare also with *Pseudosuccinea columella*.

DISTRIBUTION
This is a common Eurasian and North African species. It also occurs in Iceland. In North America it is known only from a few localities in southern Newfoundland.

ECOLOGY
Lives in ponds, other quiet waters, and stagnant habitats. In this book it is placed in *Radix* rather than *Lymnaea* because of its close relationship to *R. auricularia*.

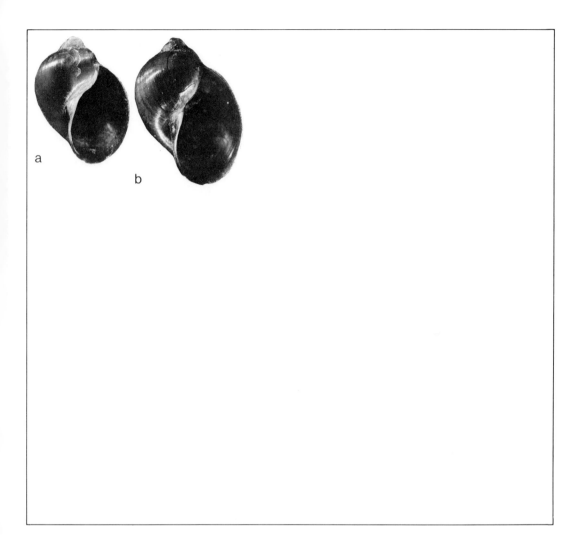

34
Radix peregra
a,b: Pond, Bellevue, Nfld.; *a* 5.2 mm, *b* 6.0 mm.

35
Pseudosuccinea columella
(Say, 1817)
American Ear Snail

DESCRIPTION
Shell up to about 21 mm high, 12 mm wide, with 4 whorls and with aperture height about 65% to 75% of shell height, succineiform, dextral, thin, fragile and with a capacious body whorl. Protoconch dark brown, small, and about 1-1/4 whorls. Spire sharply conic, rather short and narrow, and with sutures constricted and well impressed; spire angle about 50°. Whorls rounded and enlarging rapidly. Aperture large, ovate, and expanded basally. Outer lip thin and broadly curved; inner lip arched and closely appressed to the parietal wall so only a narrow slit is left open. In some specimens the inside of the empty shell is visible through the aperture almost to the apex. Columella plait spirally twisted. Periostracum light greenish brown to yellowish brown. Sculpture consists of collabral lines and streaks and spiral lines.

This medium-sized species may be recognized by its large ovate aperture and thin shell. It resembles the terrestrial snail *Succinea ovalis* but differs from that species in the possession of a distinct protoconch and a twisted columella. In living specimens the tentacles are broad and triangular and the eyes are on swellings near their bases, whereas in *S. ovalis* the eyes are at the ends of long peduncles and the tentacles are small protuberances near the bases of the eye stalks. The head-foot region is also smooth, while that of *S. ovalis* appears to be covered with scales. Fundamental internal anatomical differences also exist.

DISTRIBUTION
Occurs from southern Manitoba and the Great Lakes-St. Lawrence River system south throughout North America east of the vicinity of 100° longitude. Also occurs in Central and South America. Has been introduced into the western United States, Europe, Australia, and South Africa.

ECOLOGY
Lives in lakes, ponds, and sluggish streams among lily pads and reeds, and at the edges of these water bodies on sticks and on mud. The radula formula is approximately 35-1-35 and the lateral teeth are tricuspid.

35
Pseudosuccinea columella
a: Meach L. near Hull, Que.; 8.8 mm.
b,c: Rideau R. near Ottawa, Ont.; *b* 11.1 mm, *c* 12.7 mm.

36
Acella haldemani
(Binney, 1867)
Slender Pond Snail

DESCRIPTION
Shell up to about 25 mm high, 5 mm wide, with 5-1/2 whorls and with aperture length about 39% to 42% of shell length, slender lymnaeiform, dextral, thin, and fragile. Nuclear whorl long and ovate. Spire very narrow and long; spire angle about 20°. Whorls flat-sided, oblique, descending rapidly, and with impressed sutures that lie at 45° relative to the shell axis. Aperture long and narrow, not laterally expanded, acute above, and flared below. Outer lip thin and more or less broadly curved, and inner lip erect, straight, long, and touching (in most specimens) but not appressed to the parietal wall. Umbilicus a slender chink behind inner lip. Periostracum yellowish white to brown. Sculpture consisting of fine collabral lines and wrinkles.

The long and slender, almost needle-like shell of this species cannot be mistaken. It has a superficial resemblance to the young of *Lymnaea stagnalis* but it is much narrower and the spire sides are nearly straight, rather than concave.

DISTRIBUTION
Great Lakes-St. Lawrence drainage and vicinity in southern Ontario, southern Quebec, northern Vermont, New York, Ohio, Illinois, Wisconsin, and Minnesota.

ECOLOGY
Rare. Occurs in widely separated localities and is found only sporadically. Favourable habitats are among reeds in eutrophic lakes and ponds at a depth of about 0.3 to 1.0 metre. The radula formula is about 21-1-21 and the lateral teeth are bicuspid.

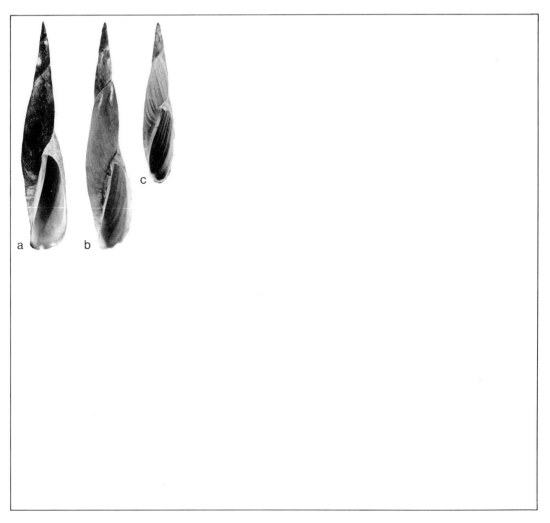

36
Acella haldemani
a,b: "New York"; 20.7 mm.
c: Georgian Bay, L. Huron, Ont.; 10.2 mm.

37
Bulimnea megasoma (Say, 1824)
Showy Pond Snail

DESCRIPTION
Shell up to about 47 mm high, 30 mm wide, with 6 whorls and with aperture height about 55% to 65% of shell height, swollen lymnaeiform, dextral, and fairly thick and solid. Nuclear whorls number 1-1/4, and are satiny and yellowish to dark brown. Spire rather short, with straight or convex sides, conical with evenly rounded whorls and impressed sutures. Body whorl dominant, large, broadly rounded, inflated but not bulbous, and comprising most of the shell volume. Aperture large, acutely angled above, flatly rounded laterally, rather sharply rounded below, and chestnut-brown to purplish within. Parietal wall flattened, with a spirally twisted columella plait, and a prominent callus that obscures the umbilicus. Periostracum glossy, brownish or greenish and with or without collabral streaks of muted green, orange, yellow-brown, or purple. Sculpture principally of coarse collabral lines and ridges.

This large, rotund, colourful species is quite different from any other. Compare with *Stagnicola catascopium*.

DISTRIBUTION
Upper Albany, Winnipeg and Nelson River drainages in the Canadian Interior Basin; Great Lakes-St. Lawrence system east to Lake Champlain; and upper tributaries of the Ohio-Mississippi River system.

ECOLOGY
Occurs in large and small lakes, in slow-moving rivers, and in pond areas of creeks. Vegetation is of variable abundance and the usual bottom is mud. Radula formulae of 48-1-47 to 49-1-49 have been observed. The lateral teeth are tricuspid.

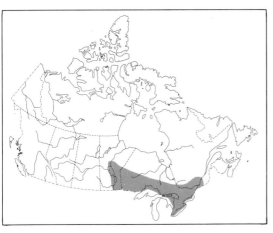

37
Bulimnea megasoma
a: Marchington R. near Drayton, Ont.; 45.0 mm.
b: Wildgoose L. near Longlac, Ont.; 37.3 mm.

38
Lymnaea stagnalis jugularis
(Say, 1817)
Great Pond Snail

DESCRIPTION

Shell up to about 56 mm high, 27 mm wide, with 7-1/2 whorls and with aperture height about 48% to 58% of shell height, lymnaeiform, dextral, and thin. Nuclear whorls smooth, shiny, satiny, forming a pointed apex and numbering about 1-1/2. Spire tall, narrow, with concave sides (in many specimens), with spire angle about 40° but variable, and with flatly-rounded whorls and impressed sutures. Body whorl capacious, inflated, shouldered in some specimens, roundly convex, and in volume constituting most of the shell. Aperture ovate except angled above; outer lip thin and fragile; inner lip with a thin but prominent callus. Columella twisted and forming a heavy, oblique spiral plait. Periostracum light tan to dark brown. Surface relatively smooth except malleated in some specimens, with prominent collabral lines, growth rests, and fine spiral lines.

This large, thin-shelled species is recognized by its long, narrow spire that has concave sides and by its bulbous body whorl. Compare it with *L. stagnalis sanctaemariae*.

L. s. appressa (Say, 1818) is a synonym. Say's first (1817) description of *jugularis* is recognizable, but that publication (Nicholson's Encyclopedia) is very rare. In later, more common editions the description was altered and became equivocal.

DISTRIBUTION

Throughout Canada south of the tree line, but absent from the region east of northern James Bay and Hudson Bay and also from the Atlantic Provinces. In the United States it occurs throughout the Great Lakes-St. Lawrence system, in the upper Mississippi River system in Ohio and Illinois, and in the Rocky Mountains south to Colorado.

ECOLOGY

Occurs in all perennial-water habitats. Vegetation is always present and bottom sediments are of diverse types; it is often found among cattails (*Typha*). Radula formulae of 37-1-36 to 46-1-46 have been recorded, and the first lateral teeth are bicuspid or tricuspid.

The nominal subspecies *wasatchensis* Hemphill, recorded from western Canada and United States, exhibits a relatively longer spire, more rounded aperture, and more inflated body whorl than typical *L. s. jugularis*. These interrelated characters are too irregularly expressed, however, to merit separate taxonomic recognition for those populations.

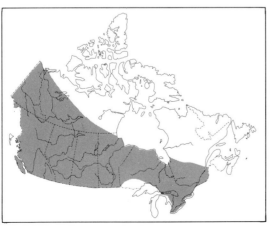

38
Lymnaea stagnalis jugularis
a: Montreal R. near Montreal L., Sask.; 50.8 mm.
b: Rideau R., Ottawa, Ont.; 61.1 mm.
c: Lake near High River, Alta.; 51.5 mm.

39
Lymnaea stagnalis sanctae-mariae Walker, 1892
Walker's Pond Snail

DESCRIPTION
Shell similar to that of *Lymnaea stagnalis jugularis* but differing principally in relative proportions. Aperture height is about 65% to 70% of the height of the whole shell, while in *L. s. jugularis* that ratio is less than 60% in nearly all specimens. Also the aperture is very wide and patulous, and the body whorl is capacious and almost globular. The spire is short, less acute, and has only about 4 whorls, whereas in *L. s. jugularis* it has 5 to 6 whorls. The last part of the body whorl in many *L. s. sanctaemariae* is also malleated or otherwise irregular.

DISTRIBUTION
Lake Superior watershed and adjacent parts of the Lake Huron, Winnipeg River and Wisconsin River watershed. Intergrades with *L. s. jugularis* occur in adjacent parts of the Nipigon, Attawapiskat, Severn, Hayes and Nelson River systems in Canada.

ECOLOGY
Characteristic of large and medium-sized lakes. Frequently found attached to large rocks exposed to wave action, but is also found in more protected habitats. Egg masses are reported as thicker and more solidly formed than in *L. s. jugularis*. The radula, however, is similar in both subspecies.

39
Lymnaea stagnalis sanctaemariae
a: Knee L., Man.; 51.7 mm.
b: Sandbar L. near Ignace, Ont.; 48.2 mm.
c: Ozhiski L., Ont.; 39.8 mm.

40
Lymnaea atkaensis Dall, 1885
Alaskan Pond Snail

DESCRIPTION
Shell up to about 42 mm high, 30 mm wide, with almost 6 whorls and with aperture height about 50% to 65% of shell height, lymnaeiform, dextral, rather thin, moderately fragile, but variable in all characteristics. Nuclear whorls rough, shining, about 1-1/2 to 2 in number, and forming a pointed apex. Spire short to long, moderately broad to narrow with an angle of 30° to 60°, scalariform, and with shouldered and strongly convex whorls. Sutures constricted and deeply impressed. Body whorl large, irregular, malleated, and shouldered in many specimens. Aperture large and ear-shaped with thin and convex outer lip and erect, continuous, reflected inner lip. Umbilicus moderately wide, deep, and revealing earlier whorls except in specimens in which it is partly obstructed by the reflected inner lip. Periostracum straw-coloured or yellowish brown. Sculpture consisting of fine collabral striae, dark-brown growth rests, spiral lines and bands, and irregular malleations.

Differs conspicuously from all other species within its geographical range except *Stagnicola catascopium*, which extends close to its range in British Columbia. That species has flatly rounded whorls, sutures that are well marked but not constricted and deep, and (in most specimens) a lower spire. The radulae and anatomy of the species are also different.

DISTRIBUTION
Lakes in Alaska (including the Aleutian Islands), the Yukon Territory, the Northwest Territories east to the vicinity of Darnley Bay (about 69°50'N, 122°E), and northern British Columbia in the Mackenzie River system (Peace and Liard River drainages) and presumably in the Yukon River system. This is a Beringian relict species.

ECOLOGY
Occurs in northern British Columbia in clear, cold, oligotrophic lakes on rocks and among sparse submersed aquatic vegetation at depths of 0.3 to 5 metres. The radula has both bicuspid and tricuspid lateral teeth, and gross formulae of 41-1-39 and 43-1-41 have been observed.

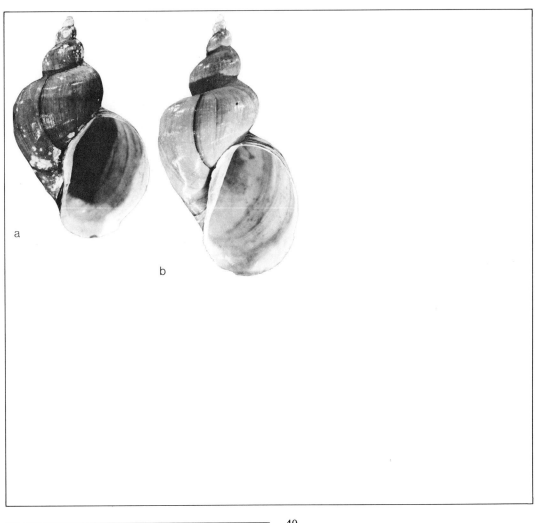

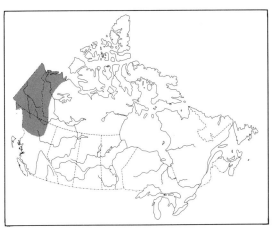

40
Lymnaea atkaensis
a,b: Lake near Paulatuk, N.W.T. (69°24′N, 124°33′W);
 a 30.1 mm, *b* 35.4 mm.

41
Stagnicola (Hinkleyia) caperata (Say, 1829)
Blade-ridged Stagnicola

DESCRIPTION
Shell up to about 16 mm high, 7 mm wide, with 6-1/2 whorls and with aperture height about 45% to 60% of shell length, lymnaeiform, and dextral. Nuclear whorls brown or reddish-brown, satiny, about 1-1/3 in number, and forming a pointed apex. Spire conoid, with flat or convex sides, with spire angle about 60°, and with whorls flatly rounded. Sutures impressed. Body whorl inflated. Aperture ovate; outer lip thin and reinforced by a varix which may be purple; inner lip broad and expanded over the umbilicus. Umbilical chink partly or wholly open. Periostracum variable, yellowish brown to blackish, and closely adherent. Most unabraded specimens (when viewed at 25× or 50× magnification) exhibit fine, low, compressed, blade-like spiral ridges standing erect in spiral grooves that are impressed in the underlying shell surface.

Recognized by its unique, microscopic, spiral, blade-like periostracal ridges. Compare with *S. montanensis* and *S. elodes*.

DISTRIBUTION
Typically a prairie species. Positively recorded from southern Manitoba to southern and central Alberta, and south in the United States to Nevada and Utah. Also reported from eastern Canada and the United States, and even from the Yukon Territory, Alaska and California, but these records are probably erroneous.

ECOLOGY
Found most frequently in temporary-water habitats (ditches, shallow pools, vernal ponds) or in spring-flooded margins of permanent-water habitats. Also occurs rarely in large permanent lakes, rivers, and swamps. Radula formulae of 28-1-28 to 35-1-32 have been observed. The lateral teeth are bicuspid.

41
Stagnicola caperata
a: Ditch near Cayley, Alta.; 10.0 mm.
b: Creek near Carstairs, Alta.; 11.0 mm.

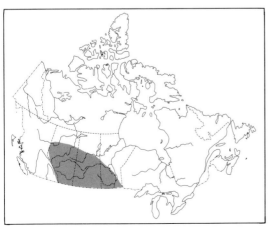

42
Stagnicola (Hinkleyia) montanensis (Baker, 1913)
Mountain-Spring Stagnicola

DESCRIPTION
Shell up to about 15 mm high, 7 mm wide, with 6-1/2 whorls and with aperture height about 38% to 50% of shell height, lymnaeiform, dextral, and of moderate thickness. Nuclear whorls brown to reddish-brown, about 1-1/3 in number, and forming a blunt apex. Spire angle about 60°; spire sides straight or bulging. Aperture relatively small, and long ovate; outer lip curved; inner lip rather straight, oblique, and sometimes angled where it is appressed to the previous whorl. Inner lip broadly reflected, sub-triangular, and exposing a small distinct umbilicus. Periostracum brownish and shiny. Under 25× magnification the sculpture is seen to consist of numerous spiral rows of tiny crescents (on most specimens) whose ends point away from the aperture, and numerous irregularly-spaced collabral lines.

Resembles *S. caperata* but differs conspicuously in its microsculpture. Compare the two species.

DISTRIBUTION
Recorded in the Rocky Mountain region from southwestern Alberta south to Nevada and Utah. May also occur in British Columbia.

ECOLOGY
Characteristic of clear, cold mountain streams and small spring-fed pools. Has also been found in a roadside ditch. A radula formula of 28-1-32 has been reported.

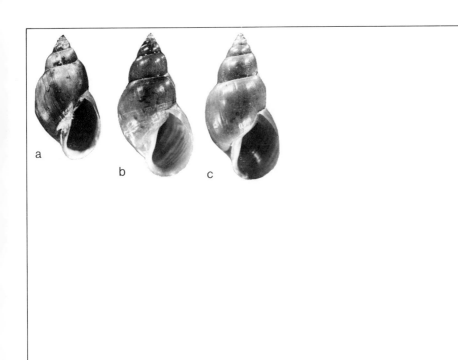

42
Stagnicola montanensis
a: Pool near High River, Alta.; 11.3 mm.
b,c: Ditch, Teton Co., Idaho; *b* 12.9 mm, *c* 13.3 mm.

43
Stagnicola (Stagnicola) arctica (Lea, 1864)
Muskeg Stagnicola

DESCRIPTION
Shell up to about 22 mm high, 11 mm wide, with 6-1/2 whorls and with aperture length about 45% to 58% of shell length, lymnaeiform, dextral, and unusually variable in all features, but normally of moderate thickness and with heavy columella development. Nuclear whorls rounded, shiny, yellowish-to-reddish brown, and about 1-1/2 in number. Spire angle about 50°; spire of medium length and with impressed sutures and convex whorls that are wider than high. Body whorl convex but not inflated. Aperture subovate and purplish brown within (in many specimens). Outer lip convex, with or without an internal varix; inner lip broad, thick, and expanded over the umbilicus, obscuring it in many specimens. Columella plait heavy, moderate, or absent. Periostracum variable, pale brown to blackish. Sculpture variable and composed of collabral lines and spiral ridges, bands, or malleations.

Resembles the widespread *Stagnicola elodes* but is smaller, with a proportionately larger body whorl and much heavier columella development.

DISTRIBUTION
An arctic and subarctic species. Occurs across northern Canada from Labrador to the Yukon Territory and beyond into Alaska.

ECOLOGY
Occurs in lakes, ponds, rivers, streams, ditches, and muskeg pools. Vegetation abundance and bottom sediments are variable. Radula formulae from 29-1-27 to 31-1-30 have been reported, and the lateral teeth are bicuspid. *Stagnicola yukonensis* Baker is a synonym.

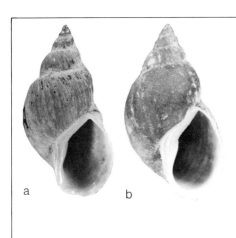

43
Stagnicola arctica
a: Moose R., Moose Factory, Ont.; 15.5 mm.
b: Hannah Bay, James Bay, Ont.; 15.7 mm.

44
Stagnicola (Stagnicola) catascopium catascopium (Say, 1817)
Lake Stagnicola

DESCRIPTION
Individual populations of this species may differ widely from each other. In some populations shells may reach 33 mm in height and 23 mm in width, but in other populations these dimensions are halved. Adult specimens have 5 to 6 whorls, and the aperture length is between 50% and 70% of shell length. Specimens may be with or without a columella plait, an open umbilicus, and strong surface sculpturing. Whorls may be rounded to nearly globose, shouldered or unshouldered, and with a flattened to pyramidal spire. The aperture is ovate to subquadrate, large to very large, and variable in degree of flare. The varix present behind the outer lip in many specimens may be brown. The inner lip is broad and partially or completely reflected over the umbilicus. Sculpturing is strong to weak, and may be composed of spiral lines, ridges and flat bands, collabral lines of growth, white varices, or any combination of these characters.

Typically a rather large and heavy shell with low spire, broad aperture, inflated whorls, deep sutures, and thickened inner lip. The only subspecies recognized as distinct are those which occur as multiple, contiguous populations and which can be differentiated by more than a single character, such as *S. catascopium nasoni* and *S. catascopium preblei*; these are discussed in the following pages.

DISTRIBUTION
Generally extends across North America below the tree line and south in the United States to about 40°N. Within these limits the only large area where the species is absent is northern British Columbia, the Yukon Territory and Alaska, that is the area occupied by *Lymnaea atkaensis*.

ECOLOGY
Characteristic of large lakes and large rivers although also found in smaller bodies of water. Its most frequent habitat is on rocks exposed to waves and currents. Radula formulae of 31-1-30 to 35-1-35 have been reported. The lateral teeth are bicuspid.

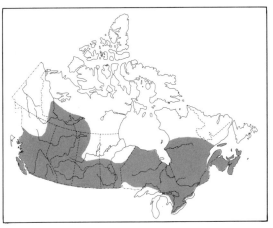

44
Stagnicola catascopium catascopium
a: Winisk R., Ont.; 18.7 mm.
b: Mistassin L., Ont.; 15.0 mm.
c: St. Lawrence R., Que.; 13.7 mm.

45
Stagnicola (Stagnicola) catascopium nasoni (Baker, 1906)
Miniature Lake-Stagnicola

DESCRIPTION
Shell up to about 12 mm high, 8 mm wide, with 4 whorls and with aperture height about 60% to 75% the height of the shell, broad lymnaeiform, rotund, solid, and with expanded aperture and broad flaring lip. Nuclear whorls dark brown, satiny, and about 1-3/4 turns. Whorls rounded but not inflated. Spire broad but acute and with impressed sutures. Body whorl subglobose and dominant. Aperture ovate, expanded, and white or brownish within; outer lip sharp but internally thickened; inner lip broad, thickened, and reflected. Umbilicus obscured by inner lip. Periostracum yellowish or purplish brown. Sculpture consisting of impressed spiral lines and fine, crowded collabral ridges.

Distinguished from *S. c. catascopium* by its unusually small size, solid appearance, and wide, flaring aperture.

DISTRIBUTION
The Great Lakes from Lake Superior to western Lake Ontario; Lake Geneva, Wisconsin; Rainy River system and Lake of the Woods.

ECOLOGY
In Lake of the Woods occurs on wave-exposed black shale rocks at, and just below, the water surface. Here the dark shells are well camouflaged and resemble air bubbles. Egg masses are about 5 to 6 mm long, crescent-shaped, and contain from 8 to 20 pale-yellow eggs. The snails move rapidly, that is like *Physa gyrina*.

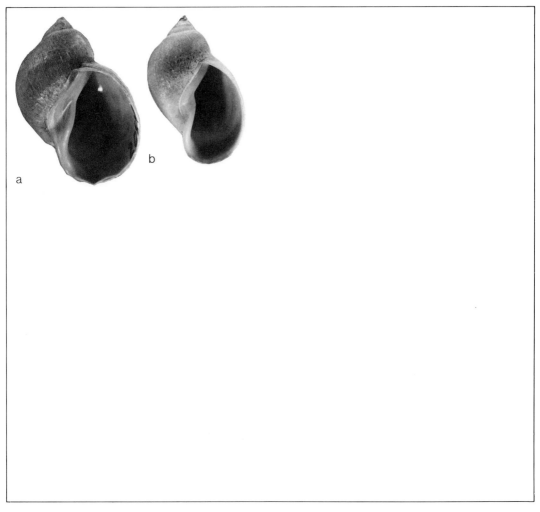

45
Stagnicola catascopium nasoni
a: L. Nipigon, Ont.; 15.4 mm.
b: Batchawana Bay, L. Superior, Ont.; 13.6 mm.

46
Stagnicola (Stagnicola) catascopium preblei (Dall, 1905)
Subarctic Lake-Stagnicola

DESCRIPTION
Shell up to 42 mm high, 28 mm wide, with almost 7 whorls and with aperture height about 55% to 65% of shell height, broad lymnaeiform, dextral, variable in shape, and with swollen whorls. Nuclear whorls brown, about 1-1/2 in number, relatively large, and forming an acute apex. Spire short but acute with deeply impressed to channelled sutures and inflated, shouldered whorls. Aperture large, ovate to subquadrate; outer lip thin and irregularly curved, with or without a reddish-brown internal varix; inner lip broad, appressed to the body whorl above and reflected over the umbilicus below but not obscuring it. Umbilicus open and deep. Periostracum pale brown. Sculpture coarse, of spiral ridges and bands, collabral lines and ridges, and irregular malleations.

Differs from *S. catascopium s. str.* in that the shell is larger, the umbilicus is entirely open rather than closed or partly open, the whorls are more uniformly shouldered, and the aperture is relatively smaller.

DISTRIBUTION
Northern Saskatchewan and northern Manitoba in the Hayes, Nelson and Churchill River systems.

ECOLOGY
Known only from large oligotrophic subarctic lakes. One specimen had a radula formula of 39-1-39 and bicuspid lateral teeth. Details of its ecology are unknown.

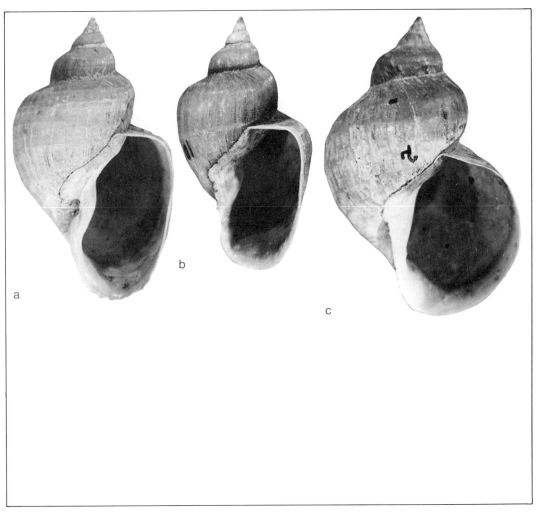

46
Stagnicola catascopium preblei
a,b,c: Limestone L., northeastern Man.; *a* 37.6 mm, *b* 33.5 mm, *c* 40.0 mm.

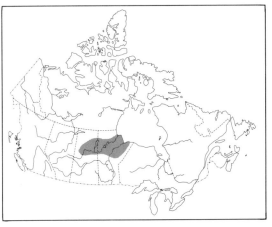

47
Stagnicola (Stagnicola) elodes (Say, 1821)
Common Stagnicola

DESCRIPTION

Shell up to about 32 mm high, 14 mm wide, with 7 whorls and with aperture height about 45% to 55% of shell height, lymnaeiform, dextral, variable in shape, and elongate. Nuclear whorls rounded, shining, and finely punctate. Spire rather long, average spire angle about 45° but variable, whorls rounded, and sutures impressed. Body whorl convex, somewhat inflated, and constituting about 2/3 the length of the shell. Aperture subovate and more or less expanded; outer lip thin, with or without an internal varix, and with a brownish-purple band within; inner lip broad and with a prominent callus. Columella plait prominent, heavy, and spiral. Umbilicus narrowly open or closed. Periostracum yellowish brown to blackish brown. Sculpturing is of numerous coarse collabral and spiral lines, a few growth rests, and (in many specimens) with spiral bands or malleations or both.

The most abundant species of its genus. In some populations specimens approach *S. proxima* or *S. reflexa* in appearance. Compare with those species.

DISTRIBUTION

Occurs throughout Canada below the tree line, and south in the United States to about 38°N. In the Rocky Mountains it extends farther south, that is to about 33°N.

ECOLOGY

Ubiquitous. Found in all kinds of aquatic habitats. Especially numerous in thick vegetation and on muddy substrates. Radula formulae of 27-1-26 to 34-1-33 have been observed.

Until recently, this species was known as *Stagnicola* (or *Lymnaea*) *palustris* (Müller).

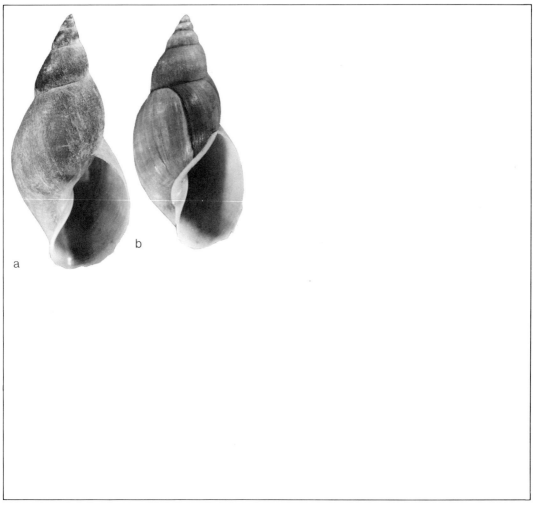

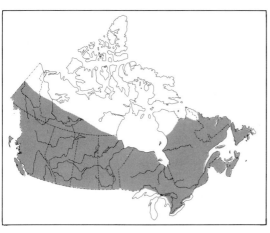

47
Stagnicola elodes
a: Pond, Quamichan, Vancouver Is., B.C.; 23.2 mm.
b: Pond, False R., Ungava Bay, Que.; 20.9 mm.

48
Stagnicola (Stagnicola) kennicotti Baker, 1933
Western Arctic Stagnicola

DESCRIPTION

Shell up to about 21 mm high, 10 mm wide, with 7 whorls and with aperture height about 35% to 45% of shell height, lymnaeiform, dextral, and solid. Nuclear whorls satiny, 1-1/2 to 2 in number, and forming a high, domed apex. Spire angle about 45°; spire rather long, with straight sides and rounded whorls. Sutures constricted, impressed and, in many specimens, incised. Aperture, roundly ovate; outer lip thin, convex, and without an internal varix; inner lip flattened, wide, reflected over the umbilicus and curved or angulate. Umbilicus visible only as a small narrow chink. Periostracum pale brown. Sculpturing is of irregular collabral lines, low spiral ridges, and spiral rows of tiny crescents between the ridges.

Related to *S. arctica* but differs in that the aperture is smaller (height less than 45% of shell height), the spire is longer, and no columella plait is present.

DISTRIBUTION

Occurs in the Northwest Territories on the mainland arctic coast of Canada from near Bathurst Inlet to Liverpool Bay, as well as on southern Victoria Island.

ECOLOGY

An arctic species. Found in large lakes, small lakes, pools, and streams. Radula formulae range from 29-1-29 to 33-1-32. The lateral teeth are bicuspid.

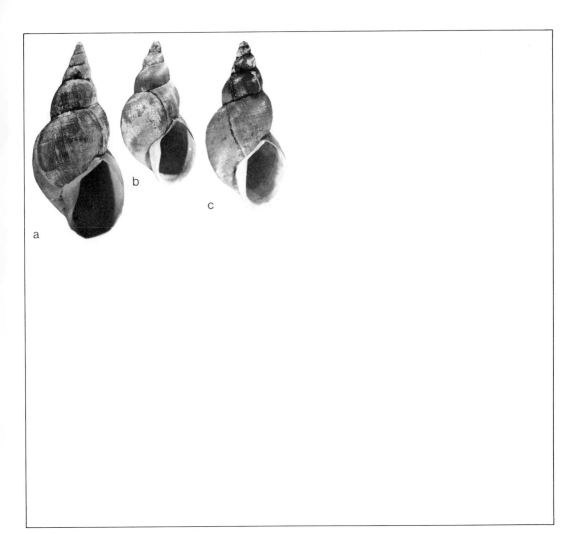

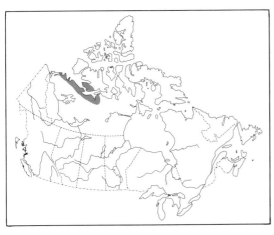

48
Stagnicola kennicotti
a: Creek, Bernard Harbour, N.W.T.; 15.1 mm (paratype).
b,c: Lake near Bathurst Inlet, N.W.T.; *b* 12.9 mm, *c* 13.6 mm.

49
Stagnicola (Stagnicola) proxima (Lea, 1856)
Rocky Mountain Stagnicola

DESCRIPTION
Shell up to about 22 mm high, 11 mm wide, with 6-1/2 whorls and with aperture height about 45% to 55% of shell height, lymnaeiform, and dextral. Nuclear whorls brown, satiny, 1-1/2 to 2 in number, and rounded. Spire sharply acute, pyramidal, pinched in many specimens, and with spire angle about 30° to 45°. Whorls flatly rounded and sutures impressed or incised. Body whorl large, well rounded to rotund. Aperture medium-sized, ovate, and (in many specimens) with a continuous lip; outer lip thin and convex; inner lip broad, reflected over the umbilicus, curved or straight, and typically without a columella plait. Umbilicus narrow but distinct in most specimens. Periostracum light brown to black. Sculpture consisting of coarse collabral lines and growth rests and less distinct spiral lines.

Resembles *S. elodes* but may be distinguished by its more rotund body whorl, rounder aperture, more sharply acute and pinched spire, and more rapidly enlarging whorls. The two taxa are closely related, however, and may be only subspecifically distinct.

DISTRIBUTION
In the Rocky Mountains from British Columbia and Alberta south to California.

ECOLOGY
A montane species. Occurs in lakes, ponds, streams, and roadside ditches. Lives on various substrates. The radula formula is 31-1-30 to 36-1-35 and the lateral teeth are bicuspid. The nominate subspecies *rowelli* appears not to be valid.

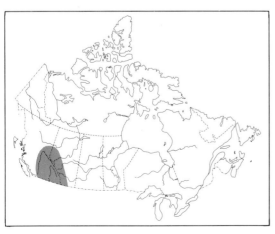

49
Stagnicola proxima
a: Maligne L., Jasper National Park, Alta.; 9.7 mm.
b: Pond, Banff, Alta.; 15.2 mm.
c: Pond, Banff, Alta.; 11.2 mm.

50
Stagnicola (Stagnicola) reflexa (Say, 1821)
Striped Stagnicola

DESCRIPTION

Shell up to about 35 mm high, 12 mm wide, with 7 whorls and with aperture height of most specimens about 44% to 50% of shell height, lymnaeiform, dextral, slender, and with flat-sided whorls. Nuclear whorls about 1-1/2 in number, turreted and satiny. Spire long, with slightly convex sides; spire angle about 25° to 30°. Sutures impressed. Last 2 or 3 whorls much elongated. Aperture long; outer lip sharp and thin but thickened internally by a brownish or reddish varix; inner lip narrow and reflected over the umbilicus and obscuring it completely or leaving only a small chink. Columella oblique and in adult specimens with a spiral plait. Surface light to dark brown, with (typically) or without alternating zebra-like dark and light collabral bands, and with numerous collabral ridges and spiral grooves.

Resembles *S. elodes* but may be distinguished by its narrower form, more-elongate later whorls, and in many populations by its zebra-like dark and light bands.

DISTRIBUTION

Scattered areas in the southern and southwestern parts of the Canadian Interior Basin, the Great Lakes-St. Lawrence system throughout, and the upper Ohio-Mississippi drainage area south to about 37°N.

ECOLOGY

Occurs among vegetation in a variety of perennial-water and vernal habitats, namely lakes, ponds, sheltered areas of streams, swamps, and ditches. The usual substrate is mud. The radula has about 40 lateral plus marginal teeth on each side of each central tooth.

Stagnicola reflexa is considered by some authors as a morphological variant of *S. elodes* (Say).

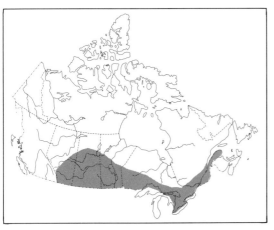

50
Stagnicola reflexa
a: L. Supérieur near Grand-Métis, Que.; 34.5 mm.
b: Pond near Starr L., Whiteshell Provincial Park, Man.; 23.0 mm.
c: Chippawa, Welland Co., Ont.; 19.7 mm.

VII Superfamily Physacea

FAMILY PHYSIDAE (Tadpole Snails)

Shell small to medium-sized, sinistral (coiled to the left), mainly thin-shelled, with an elevated spire, with or without an umbilicus and without an operculum. Tentacles slender and cylindrical, and foot narrow. The mantle in *Physa* has numerous finger-like extensions that project from the aperture over both sides of the shell. Jaw single, and radula with teeth arranged in V-shaped rows. Many multicuspid lateral and marginal teeth are on each side of the wide, multicuspid central tooth. Monoecious and capable of cross- or self-fertilization. Egg masses gelatinous, transparent, whitish, and crescent-shaped. The family is worldwide.

51
Physa gyrina gyrina Say, 1821
Tadpole Snail

DESCRIPTION

Shell up to about 24 mm high, 16 mm wide, with 5-1/2 whorls (most specimens are much smaller) and with aperture 60% to 80% of shell height, sinistral, variable in form, thin to slightly thickened, and more or less transparent. Spire acute and of medium length. Nuclear whorl small, rounded, finely punctate, and generally red to reddish brown. Whorls gently rounded and loosely coiled, that is overlapping the previous whorl only to a line that is below, or at the periphery of, that whorl. Sutures impressed and bordered below by a narrow pale band. Body whorl large and well rounded but not strongly inflated. Aperture loop-shaped, mainly basal, acute above, flatly rounded laterally, and rounded basally. Outer lip thin to slightly thickened, and bordered inside by a prominent red or reddish collabral band. Columella oblique, thin to slightly thickened, and with an extensive wash of callus on the parietal wall. Narrowly umbilicate or non-umbilicate. Periostracum pale yellowish brown to greyish brown and with a dull surface. Sculpture consists of numerous coarse lines of growth; in some specimens, one or more internal varices are visible externally as whitish collabral bands; and, in most populations, crowded, impressed spiral lines cross the growth lines.

Well characterized by its small, acute, and reddish apex; inflated whorls; and medium-sized, rather thin shell. In Canada its closest relatives are *P. heterostropha* and *P. johnsoni*; compare with these. *P. vinosa* Gould, 1847, from Lake Superior appears to be a thick-shelled ecologically induced variant of *P. gyrina*.

DISTRIBUTION

Quebec west within the tree line to the northwestern Northwest Territories, Alaska and British Columbia. South in the central and western United States to the Gulf of Mexico and California.

ECOLOGY

Abundant. Occurs in almost all perennial-water habitats and in temporarily flooded pools and swamps. Often abundant in mildly polluted water bodies; in fact, where it occurs alone and in abundance, it is indicative of organic pollution. Egg deposition occurs principally in the spring. The animal moves quite rapidly.

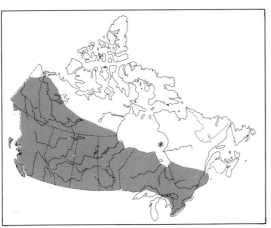

51
Physa gyrina gyrina
a: Montreal R. near Montreal L., Sask.; 18.7 mm.
b: Pond near Nevis, Alta.; 19.7 mm.
c: L. de Montigny near Val-d'Or, Que.; 11.7 mm.
d: Macamic L. near La Sarre, Que.; 13.0 mm.

52
Physa gyrina latchfordi
(Baker, 1928)
Gatineau Tadpole Snail

DESCRIPTION

Shell up to about 26 mm high, 20 mm wide, with 6-1/2 whorls and with aperture about 74% to 82% of shell height, sinistral, variable in form, slightly thickened, translucent, and shouldered. Spire obtuse, short, and with concave sides. Nuclear whorl small, reddish, and forming a sharp apex. Spire whorls flatly rounded and separated by lightly impressed sutures. Body whorl roundly but conspicuously shouldered, tabulate above, and flatly rounded below. Aperture ovate-quadrate and wider below but also rather wide above. Outer lip thin and fragile. Columella vertical, somewhat thickened, and straight or somewhat curved. No umbilicus. Periostracum yellowish brown to brown and with a shiny surface. Sculpture consists of fine collabral threads and ridges and finer spiral rows of punctae.

Distinguished by its large, wide, and strikingly shouldered shell. Intergrades occur with *P. g. gyrina*; compare with that subspecies.

DISTRIBUTION

Has been found in Meach Lake (the type locality) and Pink Lake in the Gatineau River system of western Quebec. Intergrades with *P. g. gyrina* occur in Lac Lapêche and Lac Philippe, two other lakes in the same river system.

ECOLOGY

Lives on gravel bottoms in wave-exposed areas of large lakes. Its radula is similar to that of *P. g. gyrina* except that the central tooth appears to have about 2 additional cusps. Nothing else is known about its biology or anatomy.

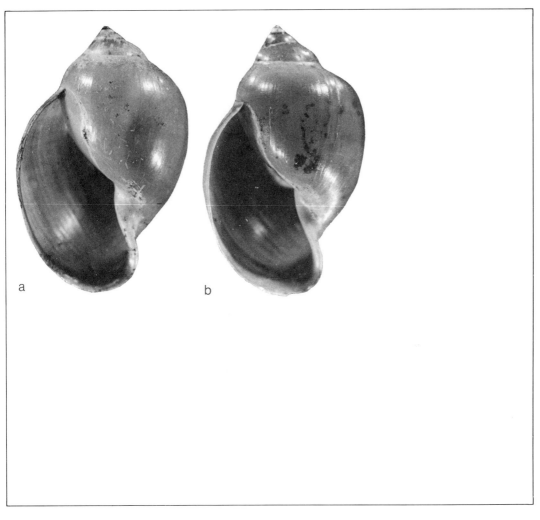

52
Physa gyrina latchfordi
a,b: Meach L. near Hull, Que.; *a* 22.0 mm, *b* 22.2 mm (topotypes).

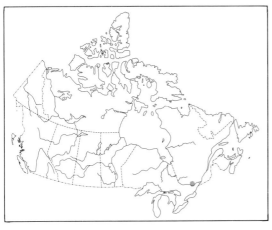

53
Physa heterostropha (Say, 1816)
Eastern Physa

DESCRIPTION
Shell up to about 21 mm high, 18 mm wide, with 6 whorls (most specimens are much smaller) and with aperture about 67% to 82% of shell height, sinistral, variable in form, thin to slightly thickened, and translucent to opaque. Spire obtuse and short. Apex decollated in most specimens, but when visible the nuclear whorl is small, rounded, finely punctate, and generally red to reddish brown. Whorls rounded or shouldered and tightly coiled, that is overlapping the previous whorl to a line that is above the periphery of that whorl. Sutures impressed and bordered below by a narrow pale band. Body whorl large and well rounded and roundly shouldered in many specimens. Aperture loop-shaped, mainly basal, acute above, rounded laterally, and sharply rounded basally. Outer lip thin to slightly thickened, and bordered inside by a prominent red or reddish collabral band. Columella somewhat oblique, thin to slightly thickened, and with an extensive callus on the parietal wall. No umbilicus. Periostracum pale yellowish brown to greyish brown, and with a shiny surface. Sculpture consists of numerous coarse lines of growth and in a few populations also of crowded, impressed spiral lines crossing the growth lines.

Similar to *P. gyrina* and, in fact, both names may turn out to apply to the same species. Typical *P. heterostropha* differs from *P. gyrina* in being more tightly coiled and in having a lower spire, a somewhat shouldered body whorl, weaker (or absent) spiral sculpturing, and shiny periostracum.

DISTRIBUTION
Atlantic Provinces of Canada south at least to New Jersey and Pennsylvania, and perhaps farther south. Its taxonomic relationships need to be clarified.

ECOLOGY
Very common. Like *P. gyrina*, occurs in all kinds of perennial-water and temporarily flooded habitats, usually among vegetation.

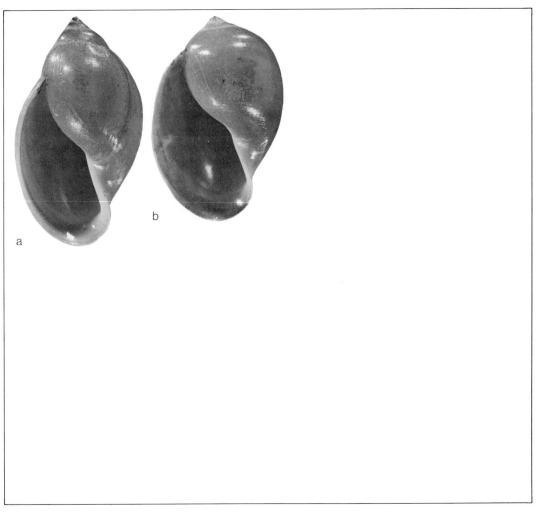

53
Physa heterostropha
a,b: Caribou Stream, Aroostook Co., Maine;
  a 13.6 mm, b 12.1 mm.

54
Physa integra Haldeman, 1841
Solid Lake Physa

DESCRIPTION

Shell up to about 15 mm high, 9 mm wide, with 5 whorls (most specimens are smaller) and with aperture about 57% to 73% of shell height, sinistral, variable in form, noticeably thickened, and opaque. Spire acute and rather short. Nuclear whorl small, rounded, finely punctate, and brown. Whorls gently rounded and loosely coiled, each overlapping the preceding whorl up to a line that is at, or below, the periphery. Sutures impressed. Body whorl large, well rounded, and somewhat shouldered in some specimens. Aperture ear-shaped, acute above, gently rounded laterally, and becoming sharply rounded basally. Outer lip thickened within by a heavy white callus (also clearly visible externally) and bordered inside by a brownish collabral band. Several whitish bands representing previous growth stages may be visible through the shell. Columella thickened and reflected over the umbilical region or forming a wide, flat expansion. No umbilicus. Periostracum yellowish brown to white, and with a dull to somewhat shiny surface. Sculpture consists of coarse lines of growth, whitish collabral bands, and (in some specimens) fine spiral lines.

Shells of *P. integra* are thicker and heavier than shells of other species; they have prominent white collabral bands and a thick, flattened inner lip.

DISTRIBUTION

Throughout the Great Lakes-St. Lawrence and Ohio-Mississippi systems.

ECOLOGY

Found in rather shallow water in lakes (but deeper than *P. gyrina*) in either exposed or protected situations, and on clay, mud, sand, or rocky bottoms. The male genitalia are distinctive (see Te 1975). The radula formula is approximately 130-1-130.

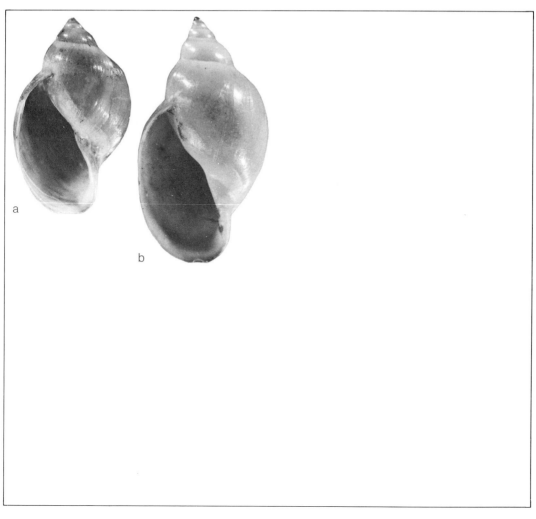

54
Physa integra
a: St. Lawrence R., Cornwall, Ont.; 11.6 mm.
b: Mohawk, N.Y.; 14.3 mm.

55
Physa jennessi jennessi Dall, 1919
Blunt Arctic Physa

DESCRIPTION
Shell up to about 9 mm high, 5 mm wide, with 5 whorls (most specimens are somewhat smaller) and with aperture 60% to 75% of shell height, sinistral, thin, and transparent. Spire acute and short with a rounded apex. Nuclear whorl large, rounded, and of the same colour as later whorls. Whorls gently rounded, the early whorls partly enveloping the preceding whorl up to a line located above the periphery, later whorls enveloping to the periphery. Spire whorls not misshapen or pulled up. Sutures shallowly impressed and bordered below by a narrow pale band. Body whorl dominant, disproportionately enlarged, roundly shouldered above, flattened laterally, sharply rounded basally, and constituting about 5/6 to 7/8 of the shell length. Aperture ear-shaped, mainly basal, acute above, flatly rounded laterally, and rounded basally. Outer lip thin and with or without a brownish collabral band within the aperture. In some specimens, broad white collabral streaks mark previous growth rests. Parietal wall covered by a rather prominent and extensive but thin callus. No umbilicus. Periostracum pale yellowish brown to greyish brown and with a shiny surface. Sculpture consists principally of collabral striae and prominent growth rests.

Differs from *P. j. skinneri* principally in that the penultimate and/or first pre-penultimate whorl is not disproportionately pulled up, while in *P. j. skinneri* that character is pronounced. The parietal callus in *P. j. jennessi* is also thicker and more prominent.

DISTRIBUTION
Known from localities close to Hudson Bay and James Bay and throughout the area between Hudson Bay and the northern Yukon Territory. Probably also occurs in Alaska.

ECOLOGY
An arctic species. Found in small lakes, ponds, woodland pools, muskeg pools, small pools on top of flat boulders, and slow-flowing streams. Vegetation present in some habitats but absent in others; substrates are rocks or mud. Its reproduction has not been investigated. The radula is similar to that of *P. j. skinneri*.

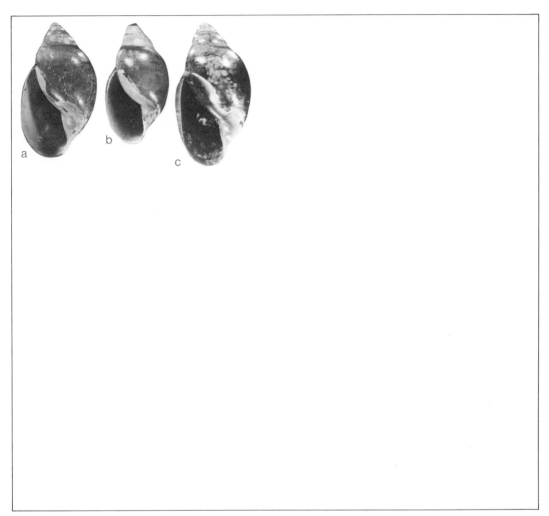

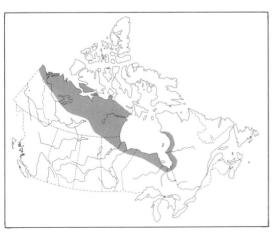

55
Physa jennessi jennessi
a,b: Creek, Bernard Harbour, N.W.T.; *a* 8.0 mm,
  b 7.0 mm (paratypes).
c: Pool, Churchill, Man.; 8.4 mm.

56
Physa jennessi athearni
Clarke, 1973
Blunt Albino Physa

DESCRIPTION
Shell up to about 7 mm high, 5 mm wide, with 4 whorls and with aperture 71% to 94% of shell height, sinistral, variable in form, subglobose, thin, and transparent. Nuclear whorl large, rounded, about 0.7 mm wide, elevated well above second whorl, and of the same colour as later whorls. Spire short, about 1/6 the length of the shell, acute to slightly obtuse, with the first 3 whorls enlarging evenly, flatly rounded, and separated by impressed sutures. Sutures bordered below by a narrow paler band. Body whorl capacious, inflated, dominant, roundly shouldered above, and flattened centrally, giving the shell a characteristic but misshapen appearance. Aperture large, pendulous, acute above, flattened laterally, and broadly rounded below. Columella sigmoid (broadly S-shaped) and parietal callus narrow above and broader below. Umbilicus entirely absent or indicated by a tiny slit. Periostracum pale yellowish white and with a shiny surface. Sculpture consisting of fine crowded collabral lines and wrinkles and, on some specimens, a single whitish narrow growth rest on the body whorl. Impressed, microscopic spiral lines that cross and interrupt the collabral wrinkles also occur on some specimens.

Differs from both *P. j. jennessi* and *P. j. skinneri* in its broader, more capacious body whorl and the whitish colour of its shell and soft parts. Also differs from *P. j. skinneri* in its shorter, more regular spire and the fewer and smaller finger-like projections on the mantle.

DISTRIBUTION
Small lakes in the Banff-Jasper region of the Rocky Mountains and in the upper Athabasca River system downstream from Jasper.

ECOLOGY
Has been recorded only from medium-sized and small lakes, and in marshes with open water. Bottom deposits are rock and gravel. The soft parts of all specimens are predominantly or wholly whitish. The radulae of specimens 6.5 mm long are about 1.5 mm in length; they have 67 to 75 rows of teeth with about 230 to 330 teeth in each row.

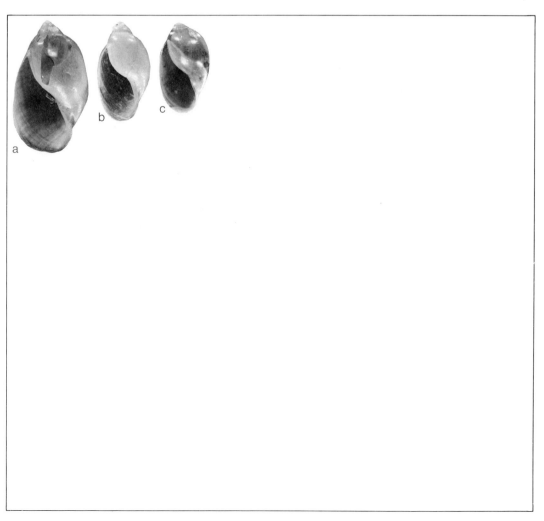

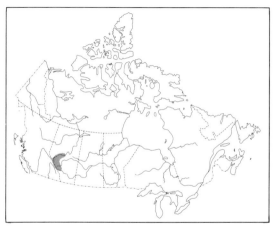

56
Physa jennessi athearni
a,b: Horseshoe L. near Jasper, Alta.; *a* 7.8 mm,
  b 5.8 mm (paratypes).
c: Marsh near Rochester, Alta.; 5.2 mm.

57
Physa jennessi skinneri
Taylor, 1953
Blunt Prairie Physa

DESCRIPTION
Shell up to about 12 mm high, 6.5 mm wide, with 5-1/3 whorls (most specimens are somewhat smaller) and with aperture about 61% to 78% of shell height, sinistral, thin, and translucent. Nuclear whorl large, rounded, finely punctate, and brown. Spire rounded and unusual, appearing peculiarly pinched and pulled-up, with the penultimate whorl, or the next preceding whorl, or both, markedly attenuate, that is much less enveloped by the succeeding whorl than are the earlier whorls and therefore more exposed. Body whorl dominant, disproportionately enlarged, shouldered above, flattened laterally, rather sharply rounded below, and constituting about 5/6 of the shell length. Aperture ear-shaped and mainly basal. Outer lip broadly curved laterally and characteristically arched centrally when viewed from the outer side. Suture weakly impressed. Parietal wall covered with a thin callus. No umbilicus. Periostracum pale yellowish brown to greyish brown and with a moderately shiny surface. Sculpture consists of collabral and spiral striae and well-marked growth rests.

The large nuclear whorl, blunt apex, and flattened body whorl distinguish *P. jennessi* and its subspecies from the *P. gyrina*-*P. heterostropha* complex. The elevated penultimate or first pre-penultimate whorl will differentiate *P. j. skinneri* from the arctic *P. j. jennessi*, with which it should be compared.

DISTRIBUTION
Southern Canada and northern United States from Ontario northwest to the vicinity of Great Slave Lake, west to British Columbia, and south to Utah.

ECOLOGY
Occurs in lakes, ponds, marshes, and slow-moving streams of all widths. Both vernal or perennial habitats are utilized. Found principally on muddy bottoms among thick or moderately thick aquatic vegetation. Recorded from depths of less than 1 m to nearly 5 m. The radula of a specimen 11.3 mm tall is 2.2 mm long and bears 71 rows of teeth with up to about 230 teeth in each row.

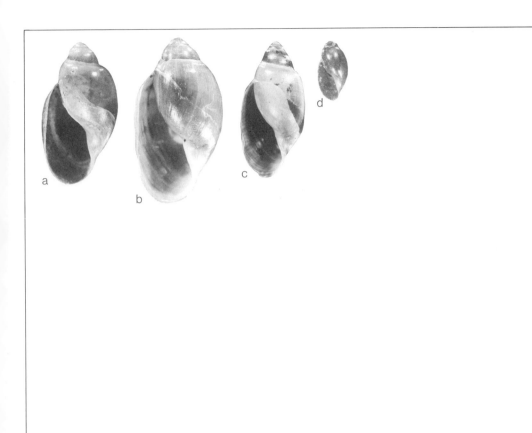

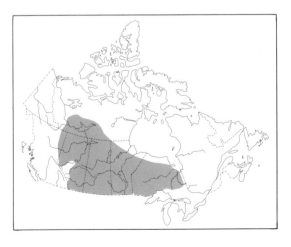

57
Physa jennessi skinneri
a: Lake near Banff, Alta.; 8.4 mm.
b: Dease L., northern B.C.; 9.7 mm.
c: Lake near Calgary, Alta.; 8.1 mm.
d: Lydia L. near Longlac, Ont.; 3.7 mm (juvenile).

58
Physa johnsoni Clench, 1926
Banff Springs Physa

DESCRIPTION
Shell up to about 9 mm high, 5 mm wide, with 4.8 whorls (most specimens are smaller) and with aperture 63% to 72% of shell height, sinistral, variable in form, thin, globose, and nearly opaque. Spire short and acute. Nuclear whorl small, rounded, and dark reddish brown. Whorls gently rounded and loosely coiled, each partly enveloping the preceding whorl to a line located at, or below, the periphery. Sutures well impressed and bordered below by a narrow pale band. Body whorl dominant, well rounded, and roundly shouldered. Aperture ear-shaped, mainly basal, acute above, flatly rounded laterally, and rounded and flared basally. Outer lip thin and without an internal dark band. Columella oblique, slightly thickened, and not terminating abruptly but continuing into the outer lip. Parietal callus very thin and not extensive. No umbilicus. Periostracum pale yellowish brown to reddish brown and with a shining surface. Sculpture consists of numerous collabral lines and in abraded specimens of irregular white collabral and spiral bands.

Differs from *P. gyrina* by its small size, lack of fine spiral striae on unabraded specimens, prominent white spiral streaks on abraded specimens, and general appearance. Quite distinct from *P. j. skinneri* as indicated by characters of the nuclear whorl, spire, and body whorl.

DISTRIBUTION
Known only from several localities in Banff National Park, Alberta.

ECOLOGY
Occurs in warm 33°C (92°F) and cold springs. Substrates are rocks and gravel with algae and moss. The radulae of two specimens, 7.6 and 6.2 mm high, were both 1.7 mm long; respectively they bore 61 and 60 rows of teeth, with about 210 and 205 teeth in each full row.

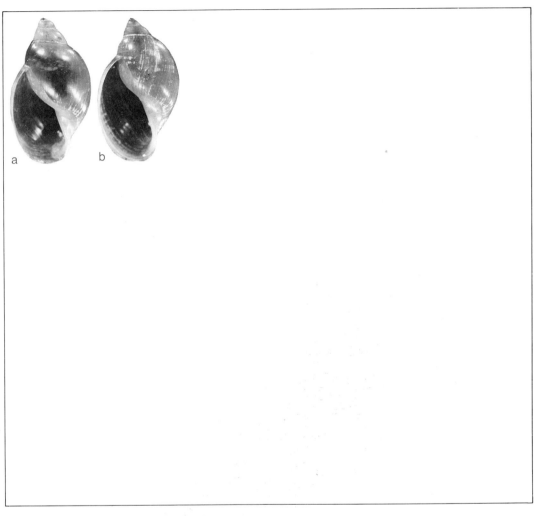

58
Physa johnsoni
a,b: Stream, Banff, Alta.; *a* 8.5 mm, *b* 8.4 mm.

59-64
Supplementary Western *Physa* Species

Six additional species of *Physa* have recently been identified from our British Columbia material by Mr. George Te at the University of Michigan; he is preparing a detailed revision of the Physidae. These species are differentiated and grouped principally on anatomical features, but useful shell characters also exist.

59. *Physa columbiana* Hemphill, 1890 (Columbia River Physa), is medium-sized, up to 15 mm high, 8 mm wide, thin to moderately heavy, with a well-marked parietal callus, flatly rounded whorls and a shiny surface. Distribution restricted to the Columbia River system. Its penial complex is similar to that of two other nominate western species, *P. hordacea* Lea and *P. lordi* Baird.

60. *Physa concolor* Haldeman, 1843 (Haldeman's Physa), is typically rather small, 7 mm high, 4.5 mm wide, 4-1/2 whorls, thin-shelled, with an extended spire, rounded whorls, and shiny surface. Has been collected near Creston, British Columbia, and in Washington and Oregon. Similar to *P. integra* Haldeman in shell characters and penial anatomy.

61. *Physa hordacea* Lea, 1864 (Vancouver Island Physa), is small, about 8 mm high, 4.5 mm wide, 4-1/4 whorls, subcylindrical, with a large nuclear whorl, blunt apex, flatly rounded spire whorls, rather narrow aperture, and shiny surface. Has been included in *Aplexa* by previous authors, but exhibits a few columellar mantle digitations, a character that is in accordance with placement in *Physa* rather than *Aplexa*. Apparently endemic to Vancouver Island.

62. *Physa lordi* Baird, 1863 (Giant Western Physa), is large, 26 mm high, 19 mm wide, thin-shelled, roundly inflated, with a pointed low spire with concave sides, an unshouldered body whorl, and a large aperture that is wider below but still rather wide above. Resembles *P. gyrina latchfordi* except that subspecies has a rather heavy shell and strongly shouldered whorls. Found from northern British Columbia (Peace River system) south to the west-central United States. *P. virginea* Gould is tentatively considered a synonym.

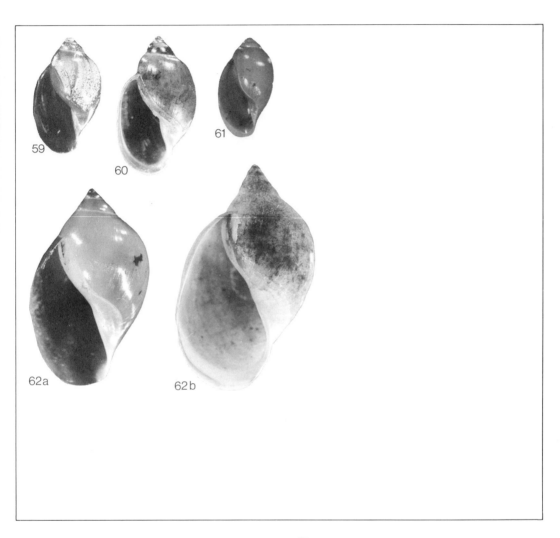

59
Physa columbiana
Christina L. near Grand Forks, B.C.; 6.6 mm.

60
Physa concolor
Moses L., Grant Co., Wash.; 8.2 mm.

61
Physa hordacea
Long L., Nanaimo, B.C.; 5.8 mm.

62
Physa lordi
a: Typical *lordi*: Vaseux L. near Okanagan Falls, B.C.; 11.9 mm.
b: Morph *virginea*: Anderson L., Jefferson Co., Wash.; 18.4 mm.

63. *Physa nuttalli* Lea, 1864 (Nuttall's Physa), is medium-sized, 12 mm high, 7 mm wide, 4-1/2 whorls, rather thin-shelled, with a large nuclear whorl, a medium-length tapering spire, flatly rounded whorls, an aperture that is wide below and narrow above, a heavy parietal callus, and with or without a brownish band within the outer lip. Occurs in ponds and ditches from southern British Columbia (near the Columbia River) to California. The penial complexes of *P. nuttalli* and *P. propinqua* are similar.

64. *Physa propinqua* Tryon, 1865 (Western Lake Physa), is rather large, 19 mm high, 13 mm wide, 5-1/2 whorls, thin-shelled, with a medium-length tapering spire, flatly rounded whorls, well-marked growth rests, a narrow and thin parietal callus, an aperture of medium width that is wider below, and with a brown band within the outer lip. Principally a lake species. Occurs on Vancouver Island and throughout central and southern British Columbia south to California.

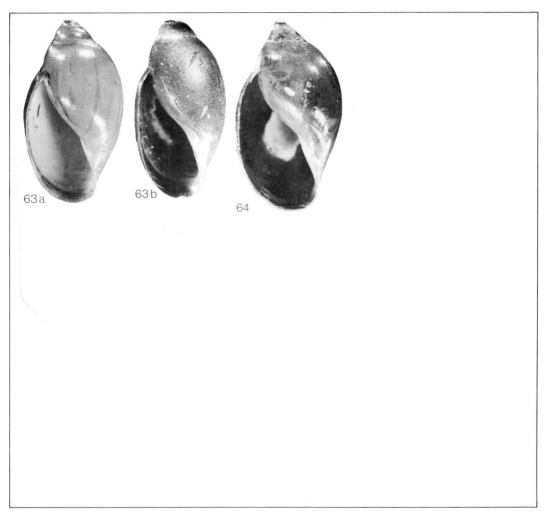

63
Physa nuttalli
a,b: Pond near Randle, Lewis Co., Wash.; *a* 10.6 mm, *b* 10.4 mm.

64
Physa propinqua
Upper Klamath L., Oreg.; 11.3 mm.

65
Aplexa hypnorum
(Linnaeus, 1758)
Polished Tadpole Snail

DESCRIPTION

Shell up to about 18 mm high, 7.5 mm wide, with 7 whorls and with aperture 50% to 60% of shell height, sinistral, elongate, thin, and transparent. Nuclear whorl rounded, finely punctate, and amber-coloured. Whorls flatly rounded and relatively tall, each partly enveloping the preceding whorl up to a line located at the periphery of that whorl. Spire elongate and constituting about 1/3 of the shell length. Sutures impressed and bordered below by a narrow white band. Body whorl flatly rounded and sub-cylindrical. Aperture acute above, rounded laterally, and flatly rounded below. Outer lip thin, flatly rounded, and only slightly thickened within. Columella oblique, narrow, slightly twisted, and with a very thin callus on the parietal wall. No umbilicus. Periostracum polished, thin, brownish (appearing blackish when containing the soft parts), and often exhibiting a greenish glint. Sculpture consists of fine lines of growth and (in some specimens, especially from arctic localities) fine spiral lines.

The long, narrow, sinistral form, polished and blackish surface (when alive), and lack of mantle digitations are characters that serve to identify this species positively.

DISTRIBUTION

New England to the District of Columbia; north to the Gulf of St. Lawrence and James Bay and across the subarctic to Victoria Island, Banks Island and arctic Alaska; and west to the Cascade Mountains. Also occurs in Europe and northern Asia.

ECOLOGY

Lives principally in vernal habitats, i.e. water bodies that dry up during parts of the year. Often very abundant in temporary shallow pools during the spring. Also found, but rarely, in large permanent rivers and lakes. Most occupied habitats have thick vegetation and a mud bottom. Adult specimens have about 350 teeth in each full radula row. Unlike *Physa*, there are no finger-like projections on the edge of the mantle.

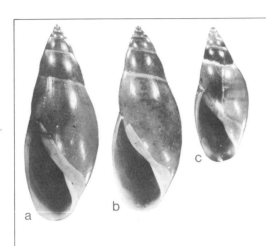

65
Aplexa hypnorum
a,b: Gatineau R., Pointe-Gatineau, Que.; *a* 16.0 mm, *b* 15.0 mm.
c: Pond near Rocky Mountain House, Alta.; 11.0 mm.

VIII Superfamily Planorbacea

FAMILY PLANORBIDAE (Ramshorn Snails)

Shells small to moderately large, dextral or sinistral, flatly coiled in most species and with a very low spire in a few, umbilicus present, and operculum absent. Tentacles long, slender, and cylindrical. An accessory breathing organ, the pseudobranch, is present and functions as a gill when atmospheric air is not available. Jaw in 3 or many segments, and radula with teeth arranged in nearly horizontal rows. About 10 to 40 tricuspid to multicuspid lateral plus marginal teeth are on either side of the small bi-, tri-, or tetracuspid central tooth. Monoecious and able to cross- or self-fertilize. Egg masses gelatinous, transparent, colourless, sausage-shaped, and with numerous yellow eggs. The family is worldwide and some tropical species are intermediate hosts for blood parasites of mammals, including man.

66
Gyraulus circumstriatus
(Tryon, 1866)
Flatly Coiled Gyraulus

DESCRIPTION
Shell up to about 5 mm wide, 1.5 mm high, with 4 whorls, planospiral, dextral, semi-transparent, and showing the soft parts within. All whorls are visible in both apical and umbilical views, and, except for the tipped-down (prosocline) aperture and collabral sculpturing, the apical and umbilical views are very similar. Whorls laterally rounded, increasing in size slowly and with outermost edge near centre. Aperture ovate, with a thin outer lip and lacking a parietal callus. Periostracum pale brown or whitish, smooth, and shiny. Sculpture consisting of fine spiral striae (in most specimens, especially on the base), collabral striae, and lines of growth.

Distinguished by being planospiral and semi-transparent and in having slowly increasing whorls and (typically) spiral striae. Compare with *G. parvus*.

DISTRIBUTION
Prince Edward Island to New England, west in the Great Lakes-St. Lawrence system, northwest to northern Alberta and British Columbia, and south in the Rocky Mountains to New Mexico.

ECOLOGY
Characteristic of small vernal habitats such as woodland pools, marshes, roadside ditches, prairie ponds, and intermittent streams. Vegetation is ordinarily thick, and substrates are commonly mud. Radula formulae of 14-1-13 to 16-1-16 have been recorded.

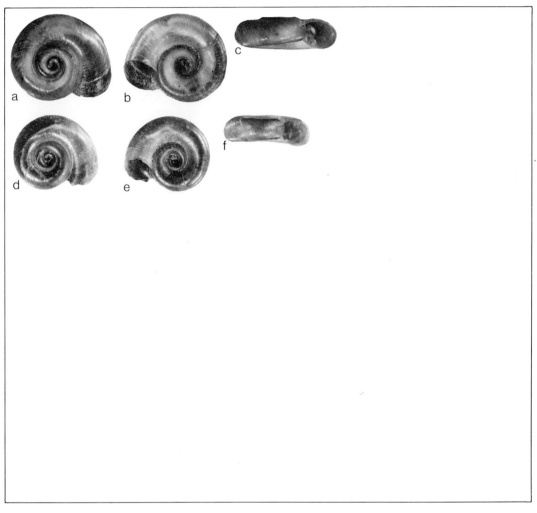

66
Gyraulus circumstriatus
a,b,c: Pond near North Portal, Sask.; 3.7 mm.
d,e,f: Another specimen, same locality; 3.1 mm.

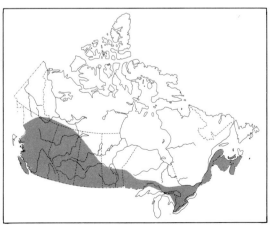

67
Gyraulus deflectus (Say, 1824)
Irregular Gyraulus

DESCRIPTION
Shell up to about 8 mm wide, 3 mm high, with 4-1/2 slowly enlarging whorls, planorboid, dextral, highly variable in form and sculpturing, and with any combination of hairy periostracum, a peripheral keel, or surface malleations. In specimens without such surface sculpturing the outer edge of the body whorl is below the midline. Apical whorls submersed. Aperture suboval, angled above, expanded at the outer edge in keeled specimens, and inclined downward. Inner lip with a thin callus. In many specimens the body whorl is also deflected downward near the aperture. Umbilicus wide, deep, and showing all whorls. Periostracum pale to dark brown and usually hirsute, that is covered with hair-like projections arranged in spiral rows parallel to the sutures. Non-hirsute specimens are spirally striate.

This small species may be distinguished from other species of *Gyraulus* by its size (it is the largest species of the genus, especially east of the Rocky Mountains), its keeled, malleated, or hirsute surface (or, if these features are absent, by the relatively low placement of the outermost edge), and by the contrasting appearance of its upper and lower surfaces. Compare with other species of *Gyraulus*.

DISTRIBUTION
Occurs throughout mainland Canada north to the central arctic, throughout Alaska, and south in the United States to Virginia and Nebraska.

ECOLOGY
Occurs in all kinds of permanent-water, eutrophic habitats. The usual substrate is mud. Commonly lives on vegetation but is occasionally found on the bottom. Egg capsules are small, gelatinous, and often contain 2 to 5 eggs. Observed radula formulae are between 15-1-15 and 19-1-18. Locomotion is rather rapid.

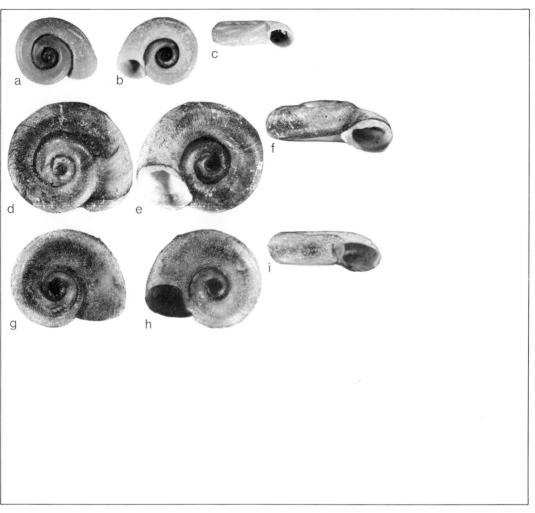

67
Gyraulus deflectus
a,b,c: Neddy Harbour Pond, Bonne Bay, Nfld.; 4.0 mm.
d,e,f: Ennadai L., N.W.T.; 6.3 mm.
g,h,i: Long L., Longlac, Ont.; 5.5 mm.

68
Gyraulus parvus (Say, 1817)
Modest Gyraulus

DESCRIPTION
Shell up to about 5 mm wide, 2 mm high, with 4 whorls, dextral, depressed but not entirely flat, and with rounded, rapidly increasing whorls. Body whorl rounded and periphery near centre. Spire flattened and first 2 whorls immersed. Aperture prosocline, ovate, in the same plane as the body whorl or somewhat below it, and with a thin rounded outer lip and a thin parietal callus. Umbilicus wide, shallow, and exhibiting all whorls. Periostracum smooth and glossy, pale brown to dark brown, and not hairy. Sculpture not prominent but limited to fine collabral striae and growth rests.

This small species may be recognized chiefly by the absence of characters that would assign it to other species, that is to *G. circumstriatus* or *G. deflectus*. It is not flatly coiled and semi-transparent, and the upper and lower aspects are clearly different. It also lacks a keel, malleations, or hirsute periostracum, and the outermost edge is near the centre of the body whorl.

DISTRIBUTION
Occurs throughout Canada and the United States south of the tree line. Also reported from Cuba and (questionably) from northern Eurasia.

ECOLOGY
An abundant species. Lives on submersed aquatic vegetation in all kinds of permanent or temporary water-filled habitats that support vegetation. Bottom most frequently muddy. Radula formulae vary from 13-1-13 to 17-1-17.

68
Gyraulus parvus
a,b,c: Peace R. near Point Providence, Alta.; 4.1 mm.
d,e,f: Herbert L. near Lake Louise, Alta.; 4.7 mm.

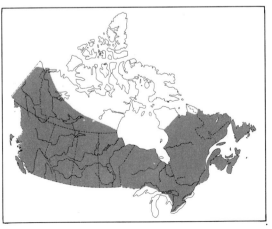

69
Gyraulus vermicularis
(Gould, 1847)
Pacific Coast Gyraulus

DESCRIPTION
Shell up to about 7 mm wide, 2.5 mm high, with about 4 slowly enlarging whorls, planorboid, dextral, and with nearly cylindrical whorls. The outermost edge of the body whorl is at the midline. Apical whorls submersed. Near the aperture the body whorl is deflected downward. Aperture elliptical and strongly slanted downward; inner lip with a thin callus. Umbilicus wide, shallow, and showing all whorls. Periostracum smooth and pale brown. Sculpture consists of low spiral threads and cords, fine crowded collabral lines, and a few irregularly spaced growth rests.

Resembles *G. deflectus*, but differs in having more cylindrical whorls and in lacking heavy malleations, a peripheral keel, or hairy periostracum. The upper and lower views are also more similar to each other than in *G. deflectus*.

DISTRIBUTION
Occurs in the Pacific coastal drainage from Alaska to California.

ECOLOGY
Lives in various kinds of perennial and vernal eutrophic habitats (lakes, ponds, sloughs, ditches, rivers, and creeks). Occurs most frequently among vegetation. Nothing has been published regarding its anatomy or life history.

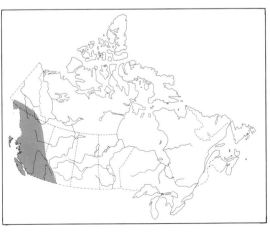

69
Gyraulus vermicularis
a,b,c: Upper Columbia L., B.C.; 6.9 mm.
d,e,f: Another specimen, same locality; 7.0 mm.

70
Armiger crista (Linnaeus, 1758)
Tiny Nautilus Snail

DESCRIPTION
Shell minute, up to about 3 mm wide, 1 mm high, with 2-1/2 whorls, dextral, thin, depressed, with a complete lip, and (in most specimens) with many prominent triangular blade-like ridges. Ridges especially elevated at the shoulder of each whorl, inclined forward, oriented parallel to lines of growth, and numbering about 12 to 18 on the body whorl. Spire flattened and sutures impressed. Whorls increasing rapidly in diameter, flattened above, rounded below, and sharply rounded at the outermost edge, which is located at the shoulder. Umbilicus wide, deep, and exhibiting all the whorls. Periostracum pale brown. In fossil specimens the periostracum is absent, and only nodular projections of the shell material are visible. Sculpture, in addition to the ridges, consisting of very fine spiral lines. Loose and partly detached coiling occurs in some specimens.

This tiny species is easily recognized by its prominent ridges, characteristic shape, and lip that surrounds the aperture.

DISTRIBUTION
Holarctic. Recorded from scattered localities in Canada from southern Ontario to Alberta and the Northwest Territories, and in the United States from Maine to Minnesota and Alaska. Also occurs in Europe, northern Asia, and North Africa.

ECOLOGY
Lives among dense vegetation in eutrophic ponds and slow-moving streams. The egg capsules are about 1.5 mm in diameter and contain 3 to 6 eggs. Radula formulae of 11-1-11 to 16-1-16 have been reported.

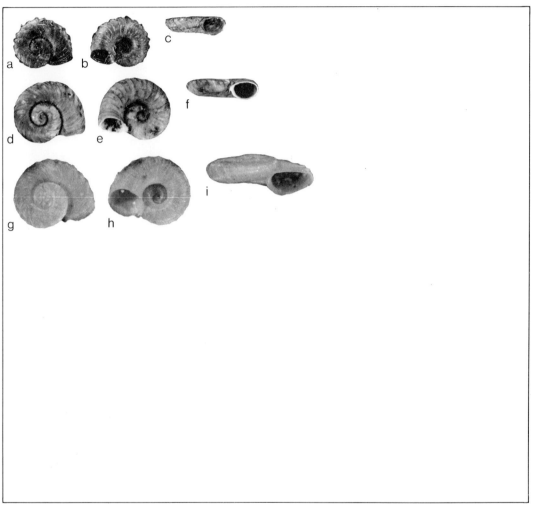

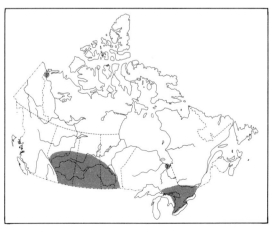

70
Armiger crista
a,b,c: Rideau R., Ottawa, Ont.; 1.5 mm.
d,e,f: L. Manitoba, Man.; 1.8 mm.
g,h,i: Marean L., northeast of Little Quill L., Sask.; 1.9 mm.

71
Promenetus exacuous exacuous
(Say, 1821)
Keeled Promenetus

DESCRIPTION

Shell up to 7 mm wide (most specimens are 6 mm wide or less), 2 mm high, with almost 4 whorls, planorbiform, biconvex, dextral, and with the outermost edge of the body whorl centrally located and prominently keeled. In most specimens the keel is sharp but in some it is rounded. Whorls expanding rapidly, broadly convex above and below, angled on the outer edge, and wider than high. Spire low, convex, or flattened. Base of aperture strongly slanted inward, almost triangular or ovate, and expanded at the outer edge. Outer lip thin to slightly thickened and inner lip with a thin callus. Umbilicus rather narrow, extending to the apex, and exhibiting all the whorls. Periostracum pale to dark brown. Collabral sculpture consisting of crowded threads and widely spaced growth rests.

Easily distinguishable from other species by its biconvex, flattened form, its prominent peripheral keel, its wide and strongly prosocline aperture, and its rapidly expanding whorls. Compare with *P. e. megas*.

DISTRIBUTION

Occurs throughout Canada south of the tree line except in Newfoundland, Labrador and subarctic Quebec. Also found in Alaska and in other parts of the United States east of the Rocky Mountains but its precise limits there are unknown. In the western prairies of Canada, it is replaced in most (but not in all) localities by *P. e. megas*.

ECOLOGY

A common species. Lives in various kinds of temporary-water and permanent-water habitats, that is large and small lakes, ponds, streams of various widths, roadside ditches, and swamps. Submersed vegetation is always present and the usual substrate is mud. Radula formulae of 16-1-16 to 18-1-18 have been recorded.

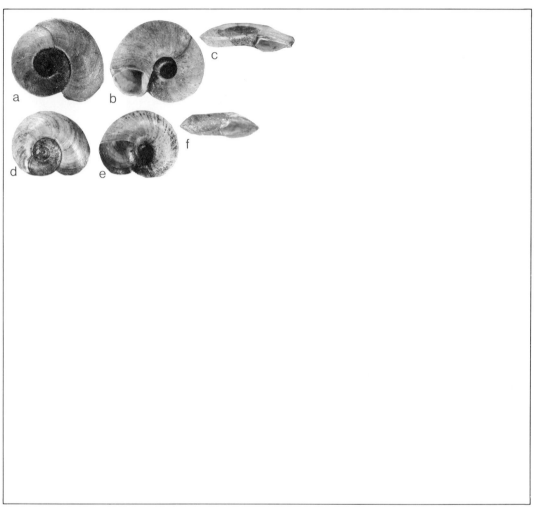

71
Promenetus exacuous exacuous
a,b,c: Pond, Moose Factory, Ont.; 4.6 mm.
d,e,f: Pond, Waterloo, Que.; 3.9 mm.

72
Promenetus exacuous megas
(Dall, 1905)
Broad Promenetus

DESCRIPTION
Similar to *P. exacuous exacuous* but its shell is larger (6-9 mm wide), heavier, and has a more prominent, pinched peripheral keel. In fresh or alcohol-preserved specimens the periostracum extends beyond the keel in a blade-like lamina. Spiral sculpture is also more apparent in *P. e. megas*, and (in many specimens) the periphery is low, causing the upper surface to be domelike and the lower to be nearly flat. The most reliable distinguishing character, however, is gigantism.

DISTRIBUTION
Typically a western prairie subspecies. Occurs from eastern Manitoba to British Columbia and in adjacent parts of the United States. Its southern limits have not been determined.

ECOLOGY
Found in lakes, ponds, streams of various widths, roadside ditches, and swamps. Most abundant where vegetation is thick and the substrate is mud. The radula formula of an Alberta specimen, 5.6 mm wide, is 17-1-17.

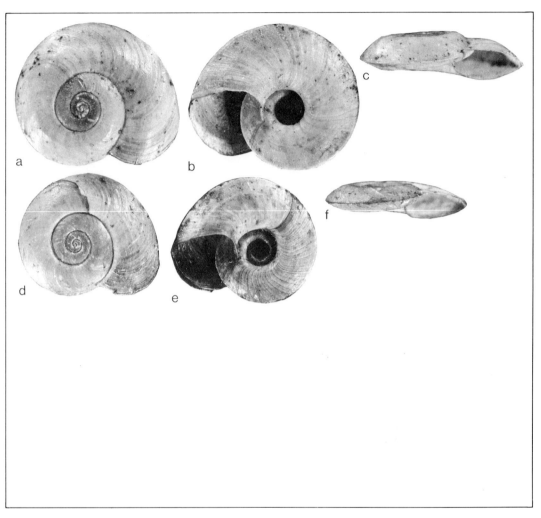

72
Promenetus exacuous megas
a,b,c: Leith R. near Whitelaw, Alta.; 8.2 mm.
d,e,f: Another specimen, same locality; 7.0 mm.

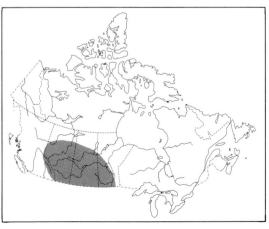

73
Promenetus umbilicatellus (Cockerell, 1887)
Umbilicate Promenetus

DESCRIPTION
Shell up to about 5 mm wide, 2 mm high, with 4 whorls, planorbiform, depressed, and dextral. Spire flattened or with the first 2 whorls slightly immersed. Whorls increasing with moderate rapidity and with regularity but somewhat expanded near the aperture. Sutures impressed. Base of body whorl flatly rounded. Aperture prosocline and roundly convex except at the inner wall, where it is abruptly concave and has a thin callus. Umbilicus prominent, wide, deep, and exhibiting all whorls. Sculpture consisting of fine collabral lines and fine revolving striae.

Distinguished by its small size, rounded whorls, and prominent, deep umbilicus. Compare with *Planorbula campestris*.

DISTRIBUTION
A western species, collected in prairie localities from southern Manitoba to central Alberta and southern British Columbia. In the United States it ranges south in the American Interior Basin to New Mexico.

ECOLOGY
Rather uncommon. Occurs in vernal ponds, marshes, and springtime flooded margins of intermittent streams. Associated with dense vegetation and mud bottoms. Radula formulae of 17-1-17 and 18-1-18 have been seen.

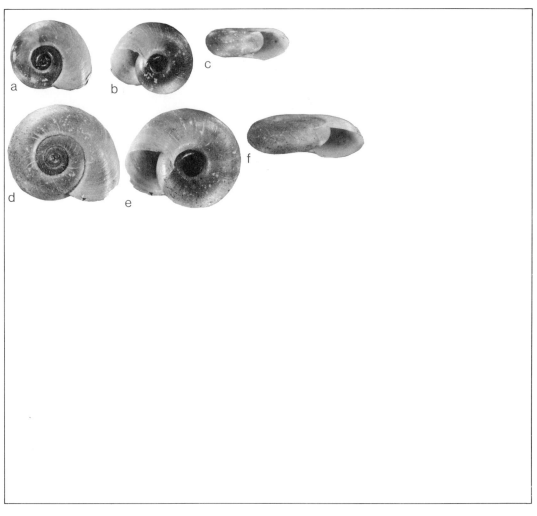

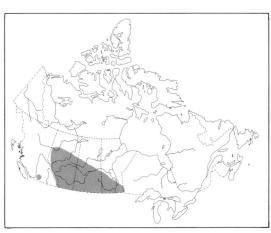

73
Promenetus umbilicatellus
a,b,c: Lake near Galahad, Alta.; 4.0 mm.
d,e,f: Another specimen, same locality; 5.6 mm.

74
Menetus cooperi Baker, 1945
British Columbia Menetus

DESCRIPTION
Shell small, up to about 8 mm wide, 3 mm high, with 4 whorls, dextral, flattened apically, convex basally, and with a carina or keel at the shoulder of the body whorl. Apical whorls somewhat immersed. Whorls expanding rapidly and separated by shallow sutures. Body whorl with a rounded angulation at the shoulder or with a carina at the shoulder that, in different populations, may be either poorly developed or prominent. Aperture prosocline, wider than high, with a thin lip and a callus deposit on the inner wall. Umbilicus broad, with a rounded angulation bounding it and exhibiting all of the whorls. Periostracum pale to dark brown, and dull to somewhat glossy. Fine sculpturing consists of crowded delicate collabral lines.

Resembles *Promenetus exacuous*, but in that species the carina is located medially on the side of the whorl, or lower, not at the shoulder. Several subspecies of *M. cooperi* and other closely related "species" have been described but require evaluation. Alberta specimens are strongly carinate, while those from coastal British Columbia localities are roundly carinate.

DISTRIBUTION
Occurs from northern California to southern Alaska and east to central Alberta. In the northern part of its range it is restricted to localities near the coast.

ECOLOGY
Found among submersed vegetation in perennial-water lakes, ponds, and slow-moving portions of rivers and streams. The radula formula is about 20-1-20.

74
Menetus cooperi
a,b,c: Whitney L. near Lindbergh, Alta.; 5.6 mm.
d,e,f: Karluk Lake, Alaska; d 5.9 mm, e 4.9 mm, f 4.1 mm

75
Planorbula armigera
(Say, 1821)
Say's Toothed Planorbid

DESCRIPTION

Shell up to about 8 mm wide, 3 mm high, with 5 whorls, planorbiform, dextral, and ordinarily with 1 set of 6 tooth-like processes deep within the aperture. Spire concave. Whorls rounded except obscurely carinate above and below and slowly expanding. Last part of body whorl expanded and abruptly deflected downward. Aperture ovate and inclined backward toward the base. Denticles located within the aperture about 1/4 whorl back and visible through the aperture and, in cleaned shells, through the shell wall. Rarely, a second set of 6 denticles is located behind the first. Umbilicus wide, deep, funnel-shaped, and showing all whorls. Periostracum pale brown to blackish. Sculpture consisting of fine lines of growth and microscopic spiral striae.

This medium-sized species (larger than most *Gyraulus* and *Promenetus* but smaller than *Helisoma*) may be easily identified by the presence of at least 1 set of 6 well-developed denticles deep within the aperture. In the western species *Planorbula campestris*, only juvenile specimens bear denticles and each set has 5 processes; the adult is also much larger.

DISTRIBUTION

In Canada this species occurs from New Brunswick to southeastern Ontario and thence northwest to the vicinity of western James Bay and beyond, within the tree line, to the north-central part of the Mackenzie River system. Also occurs in the Prairies and south in the eastern and central United States to Georgia and Louisiana. *P. jenksii* (Carpenter, 1871) is a synonym.

ECOLOGY

Lives among vegetation in most kinds of perennial-water habitats, especially stagnant, heavily-vegetated water bodies. The usual substrate is mud. It is particularly abundant in subarctic muskeg but is ordinarily rather uncommon elsewhere. Radula formulae are 18-1-18 to 24-1-24. The animal is blackish and quite active.

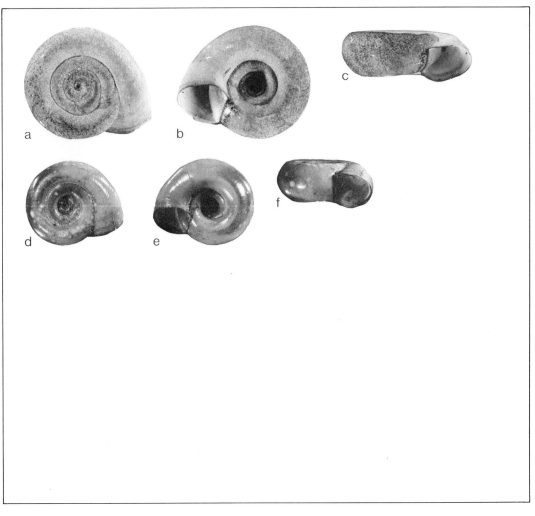

75
Planorbula armigera
a,b,c: Severn L., northern Ont.; 6.3 mm.
d,e,f: Kenogamisis L. near Geraldton, Ont.; 4.7 mm (denticles visible through shell walls).

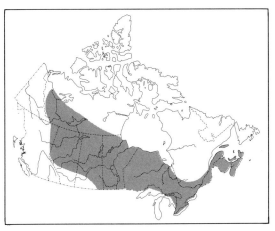

76
Planorbula campestris
(Dawson, 1875)
Prairie Toothed Planorbid

DESCRIPTION

Shell up to about 12 mm wide, 4 mm high, with 6 whorls, planorboid, dextral, and in most juveniles with 1 to 4 sets of 5 denticles arranged one behind the other within the aperture. Spire flat or centrally concave. Whorls rounded, without carinae, and increasing in size slowly. Last part of body whorl very slightly expanded and in the same plane as, or slightly lower than, the penultimate whorl. Aperture ovate except concave at inner lip. Denticles visible through the shell wall and present in most juveniles less than 4.5 mm in diameter and in some specimens up to 9 mm. In full-grown specimens denticles are absent. Umbilicus wide, deep, funnel-shaped, and exhibiting all whorls. Periostracum pale to dark brown. Sculpture consisting of fine, close collabral lines and sharp spiral lines that together produce a satin-like texture.

Juvenile specimens resemble *Promenetus umbilicatellus* except that their early whorls are much darker than the body whorl (not the case in *P. umbilicatellus*), the whorls enlarge more slowly, and in many specimens denticles are visible through the shell wall. Adults differ from *Planorbula armigera* in attaining a much larger maximum size, in having a body whorl that is not sharply deflected downward and, if denticles are present, in having 5 rather than 6 denticles in each set.

DISTRIBUTION

Principally in the western prairies but also on Vancouver Island and in southwestern Yukon Territory. Extends south in the United States to Utah and New Mexico.

ECOLOGY

Characteristic of vernal ponds, swamps, and spring-time flooded portions of permanent water bodies. Aquatic vegetation is ordinarily dense and substrates are mud. The radula of a Saskatchewan specimen 7.6 mm wide had 23 teeth on each side of the central tooth.

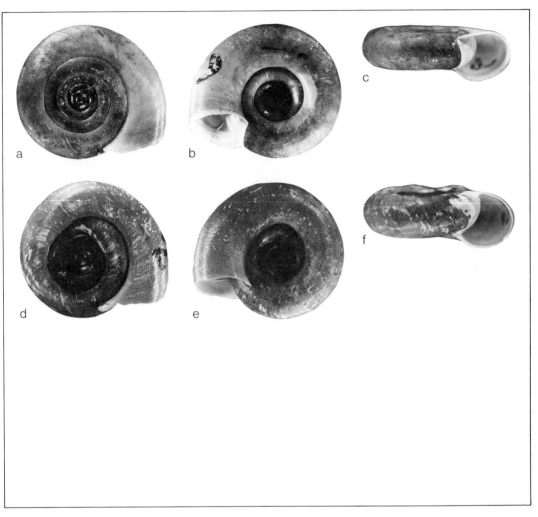

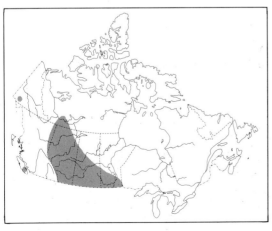

76
Planorbula campestris
a,b,c: Pond, Lintlaw, Sask.; 11.3 mm.
d,e,f: Pond near Elk Point, Alta.; 11.3 mm.

77
Helisoma (Helisoma) anceps anceps (Menke, 1830)
Two-ridged Ramshorn

DESCRIPTION
Shell up to about 20 mm wide, 12 mm high, with 4-1/2 whorls (most specimens are much smaller), planorbiform, dextral, moderately strong, and (in most populations) with a prominent carina on the upper surface of the body whorl and another around the umbilicus. Spire immersed to a variable extent but ordinarily deeply recessed. Aperture ear-shaped, expanded marginally, somewhat thickened, with a reddish band internally, and with a callus deposit on the inner wall. Carinae (where present) rounded, sharp, or corded; the upper carina located near the centre of the whorl or toward (but not on) the shoulder. Umbilicus deep and rather narrow. Periostracum pale to blackish brown. Spiral striae present on many specimens. Collabral threads fine and closely spaced.

Distinguished by its dextral coiling, medium size, and (on most specimens) prominent carinae on the upper and lower surfaces. For subspecies discrimination compare with *H. anceps royalense*.

DISTRIBUTION
Occurs throughout most of southern Canada south of the tree line. Also extends south to Georgia and northwestern Mexico, but the distributions of its subspecies in the United States have not been worked out. It has been introduced into Italy.

ECOLOGY
Lives in lakes, ponds, rivers and streams among vegetation and on various substrates. Absent from temporary-water habitats. Radulae from 2 specimens of about 10 mm diameter have 22 to 24 teeth on each side of the central tooth; in larger specimens more teeth occur. The soft parts are sinistrally oriented and the shell is therefore referred to as ultra-dextral.

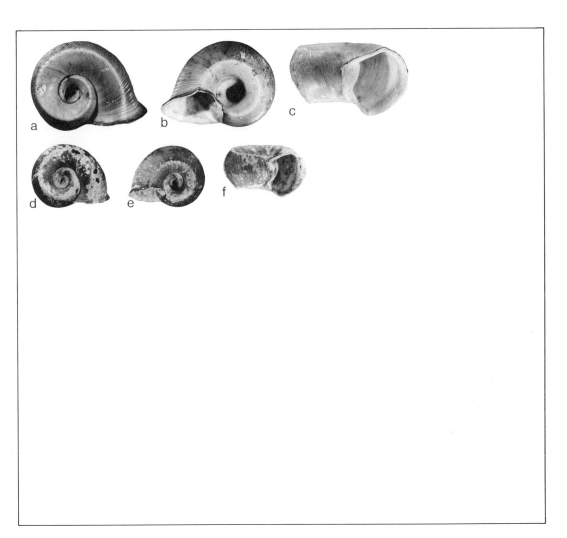
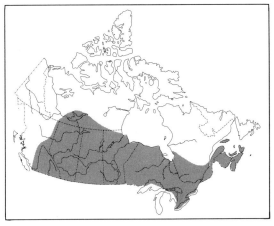

77
Helisoma anceps anceps
a,b,c: Meach L. near Hull, Que.; 20.4 mm.
d,e,f: Montreal R. near Montreal L., Sask.; 13.8 mm.

78
Helisoma (Helisoma) anceps royalense (Walker, 1909)
Lake Superior Ramshorn

DESCRIPTION
Shell up to about 17 mm wide, 10 mm high, with 4-1/2 whorls, planorbiform, dextral, moderately strong, and with a prominent carina on the shoulder of the body whorl and another around the umbilicus. Spire flat to slightly immersed. Aperture ear-shaped, marginally inflated, edged on the outside with a dark-brown band, and with an internal callus deposit on the inner wall. Upper carina sharp or rounded and forming a prominent shoulder. Body whorl laterally flattened toward the base and all whorls flattened on the upper surface. Umbilicus deep and wide. Periostracum pale brown. Collabral sculpturing typically coarse, and in many specimens one or more dark incremental growth rests are visible.

Differs from *H. anceps anceps* by having a prominent upper carina at the shoulder of the body whorl, flatter upper and basal lateral surfaces, a wider umbilicus, and (in most populations) by coarser collabral sculpturing.

DISTRIBUTION
Occurs in Lake Superior, Georgian Bay and their drainage systems, and in adjacent portions of the Albany, Attawapiskat and Winnipeg River systems in northwestern Ontario. Beyond this region a few other populations (for example, those in the Gatineau River system in Quebec) resemble this subspecies, but its principal area of occurrence is as defined above.

ECOLOGY
A distinctive subspecies. Has been collected only from lakes and large rivers. Substrates were chiefly sand or rocks; submersed vegetation was moderately dense to dense. The radula is similar to that of *H. a. anceps*.

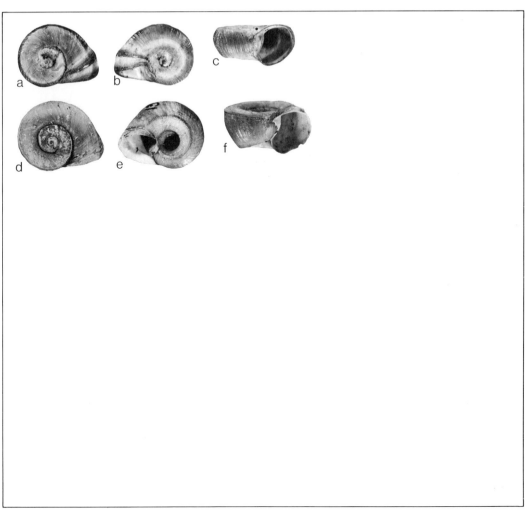

78
Helisoma anceps royalense
a,b,c: Bamaji L., Ont.; 13.8 mm.
d,e,f: Kenogamisis L. near Geraldton, Ont.; 15.0 mm.

79
*Helisoma (Planorbella)
campanulatum campanulatum*
(Say, 1821)
Bell-mouthed Ramshorn

DESCRIPTION

Shell up to about 15 mm wide, 6 mm high, with up to 7 whorls, planorbiform, sinistral, and variable in shape. Spire flattened. Whorls more or less flattened. Early whorls on upper side all visible and slightly immersed. Penultimate whorl on upper side below, or level with, body whorl, or projecting a little above it. Lower side commonly exhibiting only the ultimate and penultimate whorls. Aperture bell-shaped in profile, abruptly expanded, inverted ear-shaped in cross-section, directed slightly upward, and with a callus on the inner wall. Umbilicus narrow and deep, extending to the apex of the shell. Periostracum brownish. Spiral sculpture obscure or very fine. Collabral sculpture consisting of coarse closely spaced threads. In many specimens coiling is irregular.

The medium size, sinistral coiling, compressed whorls, flattened apex, and bell-shaped aperture of *H. campanulatum* will distinguish it from all other Canadian species. For comparison with *H. c. collinsi* see that subspecies.

DISTRIBUTION

In Canada distributed from Newfoundland to southern Quebec and west to central Saskatchewan. Two old records from the lower Fraser River in British Columbia are probably incorrect. In the United States it is found from Maine and Massachusetts to Illinois and Minnesota.

ECOLOGY

Common. Occurs in lakes and ponds of all sizes and in slow-moving or backwater portions of rivers. Vegetation is usually present and bottoms are of all types. Radula formulae are about 20-1-20 to 23-1-23. The living animal is reddish brown or blackish and is slow moving.

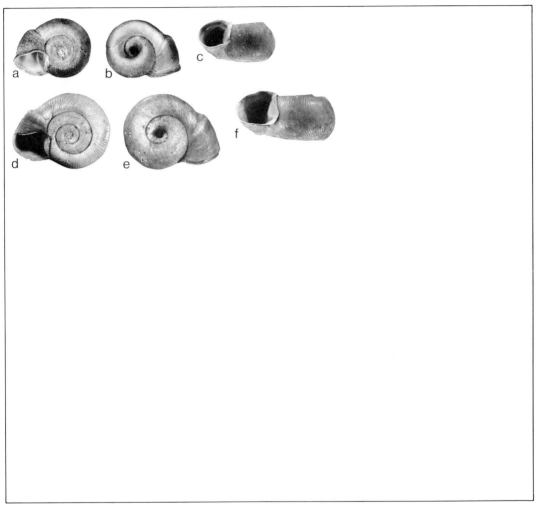

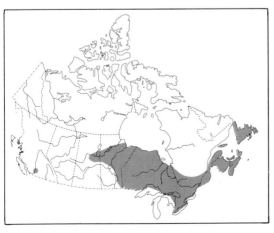

79
Helisoma campanulatum campanulatum
a,b,c: Constance Bay, Ottawa, R. near Ottawa, Ont.; 13.2 mm.
d,e,f: L. Gauvreau near Masham, Que.; 16.8 mm.

80
Helisoma (Planorbella) campanulatum collinsi Baker, 1939
Low-spired Ramshorn

DESCRIPTION
Similar to *H. c. campanulatum* except that a clearly developed spire is present. Populations of individuals in which earlier whorls project above the body whorl an average of 10% or more of the height of the body whorl are considered to be *H. c. collinsi*.

DISTRIBUTION
Lake Superior, its drainage system, and in headwaters of the Albany, Winnipeg and Severn River systems. A related form, called *H. multivolvis* (Case), occurs near Lake Superior in northern Michigan and is probably only an extreme morph of *H. c. collinsi*.

ECOLOGY
Has been collected in lakes and in medium-sized rivers. Substrates are sand or sand and gravel; vegetation is present but varies in abundance. The radula formula of a specimen 12 mm wide was 22-1-21.

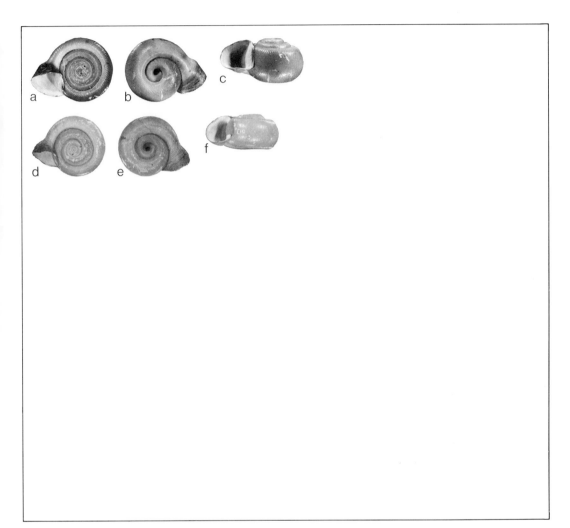

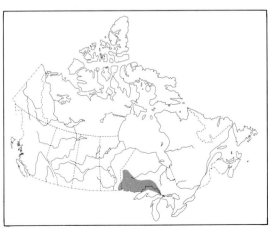

80
Helisoma campanulatum collinsi
a,b,c: Wabaskang L. near Vermilion Bay, Ont.; 14.0 mm.
d,e,f: North Spirit L. at outlet, Northern Ont.; 12.9 mm.

81
Helisoma (Pierosoma) corpulentum corpulentum (Say, 1824)
Capacious Manitoba Ramshorn

DESCRIPTION
Shell up to 32 mm wide, 16 mm high, with 4-1/2 whorls, planorboid, sinistral, and with a carina located near the outer edge of each whorl both above and below. Upper side flat centrally or slightly concave except for the body whorl, which protrudes above earlier whorls. Carinae on upper surfaces of whorls sharp and forming a 90° angle at the shoulder of early whorls but becoming rounded and more centrally located on the body whorl of mature specimens. Umbilical carina forming an acute angle on the outer edge of early whorls but also becoming rounded and central on the body whorl. Body whorl flattened and in some specimens bent upward near the aperture. Aperture large, dilated, higher than wide, and extending above, or above and below, the body whorl. Umbilicus wide, deep, and extending to the apex. Periostracum light to dark brown. Fine sculpture consisting of prominent, crowded, strongly elevated collabral riblets and microscopic collabral and spiral striae.

Distinguished by its large and high form, sharp carinae on the shoulders, laterally flattened whorls, and coarse sculpturing. Compare with *H. c. whiteavesi*, *H. pilsbryi infracarinatum*, and *H. trivolvis trivolvis*.

DISTRIBUTION
Northwest and west of Lake Superior in the Winnipeg, upper Albany and upper Severn River systems, and in the upper Mississippi River system in northern Minnesota.

ECOLOGY
Lives in large and small lakes and in rivers, and often in exposed habitats. Vegetation may be sparse to thick, and substrates are of all kinds. Diverse radula formulae between 27-1-26 and 45-1-45 have been recorded.

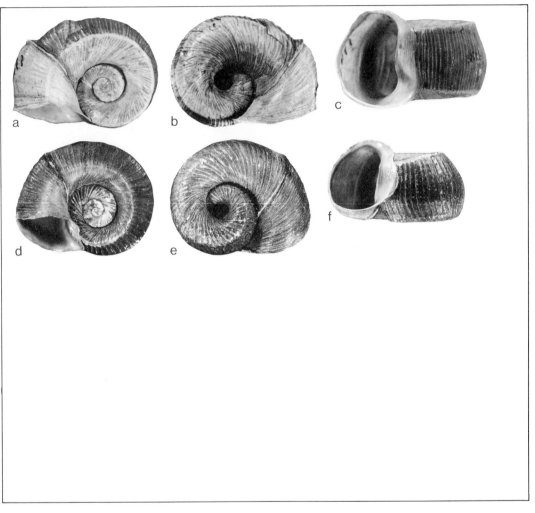

81
Helisoma corpulentum corpulentum
a,b,c: L. Seul, northwestern Ont.; 25.8 mm.
d,e,f: L. la Croix, Rainy River Dist., Ont.; 23.6 mm.

82
Helisoma (Pierosoma) corpulentum whiteavesi Baker, 1932
Whiteaves's Capacious Ramshorn

DESCRIPTION
Similar to *H. c. corpulentum* except that it is relatively higher (the holotype is 24 mm in diameter and 19 mm high), the upper side is flatter, and the whorls are more tightly coiled. The large ear-shaped aperture gives the shell a *Physa*-like appearance.

DISTRIBUTION
Known only from Lac des Mille Lacs (the type locality) and Greenwater Lake, both northwest of Lake Superior in the Winnipeg River system in northwestern Ontario.

ECOLOGY
This problematic and rare subspecies apparently occurs in open-water habitats in lakes. The radula formula is reported as 36-1-36 to 42-1-42.

Another subspecies, *H. c. vermilionense* Baker, 1929, is recorded from the Rainy River system in northern Minnesota and may occur in Canada. See Clarke (1973) for additional details regarding both of these subspecies.

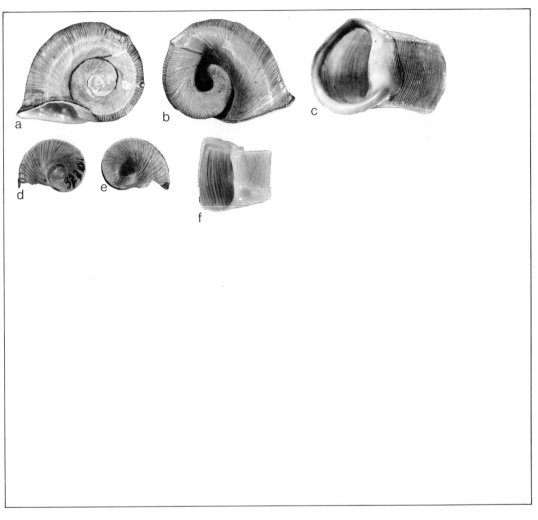

82
Helisoma corpulentum whiteavesi
a,b,c: Greenwater L. west of L. Superior, Thunder Bay Dist., Ont.; 23.5 mm.

Helisoma corpulentum vermilionense
d,e,f: Vermilion L., St. Louis Co., Minn.; 12.2 mm (juvenile, paratype).

83
Helisoma (Pierosoma) pilsbryi infracarinatum Baker, 1932
Greater Carinate Ramshorn

DESCRIPTION
Shell up to 32 mm wide, 18 mm high, with 4-3/4 whorls, planorbiform, sinistral, and with sharp or rounded carinae above and below. These carinae are more or less centrally located but each may be rounded and obsolete on the body whorl. Body whorl overlapping the penultimate whorl, and in many specimens the upper side of the shell is smoothly concave and bowl-like. Body whorl rounded on its outer side but with the outermost edge below the centre. Aperture ovate ear-shaped, higher than wide, and flaring. Umbilicus wide and exhibiting 2 to 3 whorls. Periostracum yellowish to brownish. Collabral sculpture consisting of fine riblets (2 to 4 per mm) and, in some specimens, of one or more growth rests.

Differs from *H. trivolvis* in the possession of a basal carina and in greater axial height. May be distinguished from the *H. corpulentum* group by its bowl-like (not flattened) spire, by the carinae that are more or less centrally located (not near the outer edges of the whorls), and by the more sharply convex lateral area of the body whorl.

DISTRIBUTION
In the boreal forest from southwestern Quebec to east-central Alberta (Lac la Biche) and farther south in the Rideau and St. Lawrence rivers; in Georgian Bay; and in the Qu'Appelle River system in Saskatchewan.

ECOLOGY
Ordinarily occurs in lakes, ponds, or quiet backwaters of streams, among vegetation, and on various substrates. The radula formula is approximately 30-1-30 to 37-1-37.

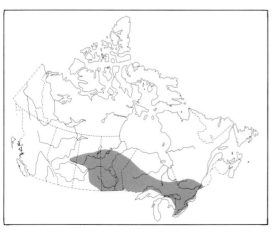

83
Helisoma pilsbryi infracarinatum
a,b,c: Rapids on Basswood R. west of L. Superior, Ont.; 23.0 mm (paratype).
d,e,f: Knee L. at outlet, northern Man.; 28.0 mm.

84
Helisoma (Pierosoma) trivolvis trivolvis (Say, 1816)
Larger Eastern Ramshorn

DESCRIPTION
Shell up to about 32 mm wide, 16 mm high, with 5 whorls, planorbiform, and sinistral. Spire submersed and dished. Whorls more or less carinate centrally or rounded above, rounded below, and rounded on the outer side. Apical depression wide, of moderate depth, and exhibiting all whorls. Umbilical region sunken and revealing 3 to 3-1/2 whorls. Aperture expanded, dilated at the margin, ovate ear-shaped, with width equal to or greater than height, with a callus deposit on the inner wall, and typically with a reddish-brown or purplish band within the aperture. Umbilicus broad and deep. Periostracum yellowish brown to brown. Sculpture moderate and consisting of collabral riblets (about 2 to 4 per mm), irregularly spaced growth rests, and in many specimens a more or less apparent spiral carina above.

This is the most abundant of the large eastern helisomas. The close-to-centre position of the upper carina will distinguish it from the *H. corpulentum* group, in which the carina is at the shoulder. The lack of a lower carina will differentiate it from *H. pilsbryi infracarinatum*, which has a well-developed lower carina. Its substantial axial height (more than 12 mm) will separate it from the western *H. trivolvis subcrenatum* and from the other species of the genus. Compare *H. t. binneyi*.

DISTRIBUTION
In Canada occurs throughout the boreal and deciduous forest regions from eastern Quebec and Nova Scotia to southeastern Manitoba and within a small area in central Saskatchewan. Also extends throughout the northeastern United States and farther south, where it intersects other subspecies, whose precise distributional limits have not been established.

ECOLOGY
Characteristic of well-vegetated perennial-water lakes, ponds, and slow-moving streams. Mud is the usual substrate. Radulae from Ontario specimens 25 mm in diameter have about 31 to 34 teeth on each side of the central tooth.

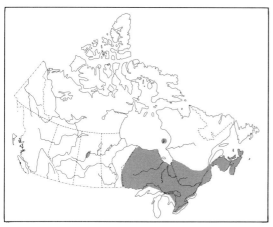

84
Helisoma trivolvis trivolvis
a,b,c: Montreal R. near Montreal L., Sask.; 26.8 mm.
d,e,f: Lake on Green R., N.B.; 23.7 mm.

85
Helisoma (Pierosoma) trivolvis binneyi (Tryon, 1867)
Binney's Stout Ramshorn

DESCRIPTION

Resembles *H. t. trivolvis* except for the following: (1) it is relatively higher, that is within population samples the average ratio of height to diameter, excluding the lip, exceeds 0.57, whereas in *H. t. trivolvis* it seldom exceeds 0.52; (2) a more or less well-defined carina occurs on the undersides of the whorls but not in *H. t. trivolvis*; (3) the body whorl is more expansive, causing the apex and the umbilicus to be more deeply immersed; and (4) the aperture is more broadly expanded above and below.

DISTRIBUTION

Principally in the Pacific drainage from southern British Columbia to California but also in a few lakes in western Alberta. In Lake Wabamun, Alberta, intergrades completely with *H. t. subcrenatum*. Populations intermediate between *H. t. binneyi* and *H. t. subcrenatum*—corresponding to "*H. binneyi randolphi*" Baker, 1945, and "*H. columbiense*" Baker, 1945—also occur in British Columbia. Herein, *H. t. binneyi* is used only for the extreme morph described above. Intermediate populations are best referred to as *H. trivolvis* Say, without any subspecies name.

ECOLOGY

Occurs in eutrophic, well-vegetated lakes. No detailed information is available regarding its specific ecology or anatomy.

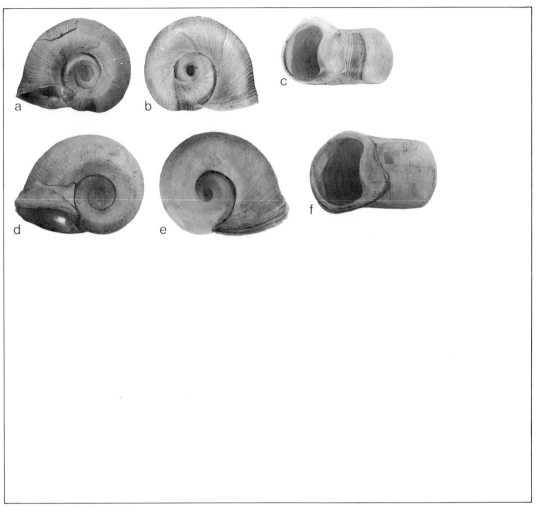

85
Helisoma trivolvis binneyi
a,b,c: Dragon L. near Quesnel, B.C.; 20.5 mm.
d,e,f: East end of Sylvan L., Alta.; 22.7 mm.

86
Helisoma (Pierosoma) trivolvis subcrenatum (Carpenter, 1856)
Larger Prairie Ramshorn

DESCRIPTION
Similar to *H. t. trivolvis* except that its height does not exceed 10 mm whereas the height of adult *H. t. trivolvis* almost always exceeds 10 mm and in most specimens exceeds 12 mm; also its coiling is looser and sutures deeper. Some populations exhibit spiral, pale-coloured streaks or irregular coiling. The two subspecies occupy distinct zoogeographical regions.

DISTRIBUTION
Western North America from California and Utah to Yukon Territory and Manitoba.

ECOLOGY
Occurs in nearly all perennial-water habitats that support significant rooted vegetation. Mud is the most frequent substrate. Specimens from Alberta about 20 mm wide have radulae with 28 to 30 teeth on each side of the central tooth.

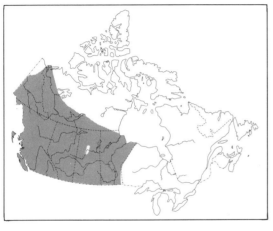

86
Helisoma trivolvis subcrenatum
a,b,c: Loch Haven, Cypress Hills Provincial Park, Sask.; 20.8 mm.
d,e,f: Third Vermilion L., west of Banff, Alta.; 26.8 mm.

FAMILY ANCYLIDAE
(True Freshwater Limpets)

Shell small to small-medium, not spiral but shaped like a broad, low cone, mainly thin-shelled, with rounded or ovate aperture, radial and concentric sculpturing, and a blunt apex located either centrally or behind the centre and either in the midline or to the right. The soft parts are sinistrally organized, with a pseudobranch (false gill) on the left side. The radula has about 8 to 37 bicuspid to multicuspid lateral plus marginal teeth on each side of the bi- to tetracuspid central tooth. The egg masses are small, ovate, gelatinous, transparent and have a few whitish eggs. The family is worldwide.

87
Laevapex fuscus
(C.B. Adams, 1841)
Dusky Lily-Pad Limpet

DESCRIPTION

Shell up to about 8 mm long, 5 mm wide, 3 mm high (most specimens are much smaller), broad cone-shaped, thin-shelled, more-or-less elliptical at the base and rather low. Apex elevated, sharply rounded, located to the right of the midline and posterior of centre, and without microscopic radial lines. Anterior slope slightly convex, posterior slope concave, and lateral slopes more or less straight. Aperture elliptical or with sides a little flattened and convergent posteriorly. Concentric lines of growth very fine. Radial lines also seen on the sides of some specimens below the apex. Periostracum very thin, closely adherent, and pale brown.

    Distinguished from *Ferrissia* species by having a smooth rather than radially striate apex and by characters of the soft parts, which are visible through the shell, that is the presence of a dark pigment band across the middle of the mantle and the presence of epithelial attachment areas between the anterior muscle scars and between the posterior and the right anterior muscle scars. These characters are best seen on specimens that are well cleaned and have had the periostracum removed by soaking in a weak solution of laundry bleach in water.

DISTRIBUTION

Southern Ontario and Quebec to Massachusetts, south to Florida and Louisiana, and west to Oklahoma, Kansas, and Iowa. Also in the upper Albany River system in northwestern Ontario.

ECOLOGY

Occurs in heavily vegetated permanent-water habitats on the undersides of lily pads and on other emergent vegetation. The radula is similar to that of *Ferrissia* species but the genitalia are distinctive, that is the penis is large and without a flagellum; in *Ferrissia* the penis, when present, is small and has a club-shaped flagellum.

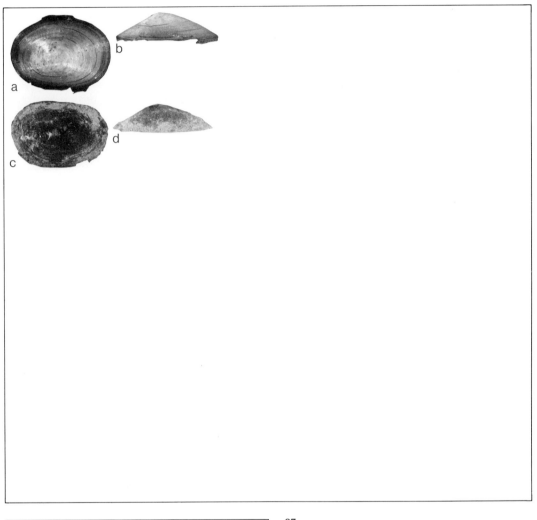

87
Laevapex fuscus
a,b: Sydenham R., Loughborough Twp., Ont.; 4.9 mm.
c,d: Carp R., Carp, Ont.; 4.7 mm.

88
Ferrissia fragilis (Tryon, 1863)
Oval Lake-Limpet

DESCRIPTION
Shell up to about 5.5 mm long, 3.2 mm wide, 1.6 mm high (most specimens are much smaller), broad cone-shaped, thin, subovate, wider anteriorly than posteriorly, low, and highly variable in form. Apex elevated, rounded, located in the midline (or a little to the right) behind the centre, and sculptured with numerous very fine radial striae that are visible at about 50× magnification. Anterior slope convex, posterior slope concave, and lateral slopes straight. Aperture suboval in one plane, abruptly rounded anteriorly and posteriorly, and flatly convex laterally. Lateral margins converging posteriorly in many specimens, causing the posterior margin to be more sharply curved than the anterior. Periostracum very thin, closely adherent, and pale brown. Concentric lines of growth very fine, and radial sculpture poorly marked.

Distinguished by its small size (less than 4 mm long in most specimens), thin shell, rather straight sides that typically converge posteriorly (in western Canadian specimens), and its predominantly still-water habitat. See Basch (1963) for a discussion of the variation in this limpet.

DISTRIBUTION
Common in southwestern British Columbia and in southern Ontario and Quebec. Widely distributed in the United States.

ECOLOGY
Occurs principally in lakes, ponds, and ditches. Often found on the stems of cattails (*Typha*), near their bases. Also lives in slow-flowing streams. The radula is similar to that of other species of *Ferrissia*. A septate morph resembling the marine genus *Crepidula* (slipper limpets) exists in some localities in the United States.

Basch (1963) considers the morph occurring from British Columbia to Oregon to be distinctive, and has called it "form" *isabellae*. It is larger than typical *F. fragilis*, and approaches *F. parallela* in shell appearance.

88
Ferrissia fragilis
a,b: Cowichan L., Vancouver Is., B.C.; 4.9 mm.
c,d: Another specimen, same locality; 5.4 mm.

89
Ferrissia parallela
(Haldeman, 1841)
Flat-sided Lake Limpet

DESCRIPTION
Shell up to about 7.6 mm long, 3.2 mm wide, 2.7 mm high, limpet-shaped, thin-shelled, long, narrow, and with aperture sides straight and parallel or somewhat convergent anteriorly. Apex elevated, obtuse, located in the midline slightly behind the centre and covered with very fine radial striae, visible at 50× and preserved principally on young specimens. Left and right lateral surfaces flattened, posterior slightly concave, and anterior a little convex. Aperture long, ovate, and with lip all in the same plane or slightly saddle-shaped, i.e. concave at the ends and convex at the sides. Periostracum very thin, closely adherent, and yellowish brown. Lines of growth mostly fine and well marked. Radial striae on body of shell obscure but generally distributed.

Easily distinguished by its long, narrow shell, its straight, more or less parallel sides, and its habitat.

DISTRIBUTION
Occurs in southern Canada from Newfoundland and Prince Edward Island to southern Manitoba. According to the literature (Basch 1963), it is also found across the northern United States.

ECOLOGY
Lives in lakes, swamps, and slow-flowing rivers among thick or moderately thick vegetation. Often found on stems of cattails (*Typha*) and sedges (*Scirpus*) or on the undersides of lily pads. Egg capsules are white and most contain 1 to 3 eggs. The central tooth of the radula has 2 large central cusps and 2 small marginal ones; the lateral and marginal teeth are multicuspid.

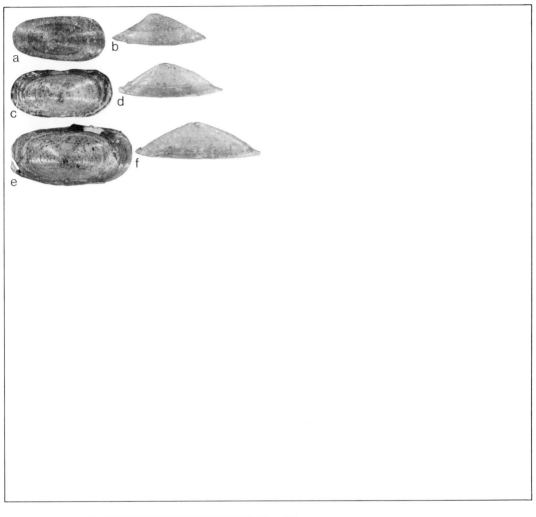

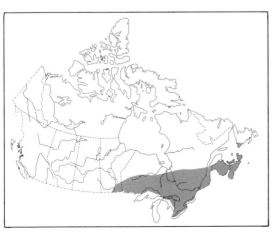

89
Ferrissia parallela
a,b: Klotz L. near Longlac, Ont.; 4.6 mm.
c,d: Carp R. near Carp, Ont.; 5.1 mm.
e,f: Another specimen, same locality; 6.0 mm.

90
Ferrissia rivularis (Say, 1817)
Sturdy River Limpet

DESCRIPTION
Shell up to about 7 mm long, 4 mm wide, 3 mm high, limpet-shaped, thin to noticeably thickened, ovate, high or low spired, and variable in form. Apex elevated, rounded, located in the midline (or a little to the right) behind the centre, and covered with very fine radial striae. Striae preserved best on young specimens and visible at $50\times$. Anterior slope convex, posterior concave, and lateral slopes more or less straight. Aperture oval with all margins convex and in the same flat plane. Periostracum thin, closely adherent, and pale brown. Lines of growth mostly fine and irregular, radial sculpturing obscure and principally anterior.

Distinguished by its oval and often thickened shell, by its relatively large size, and by its habitat.

DISTRIBUTION
Eastern Canada from New Brunswick to Saskatchewan. In the United States at least from Maine to North Dakota, and North Carolina to Wyoming.

ECOLOGY
Lives attached to rocks and mussel shells in rivers and creeks, or attached to rocks in exposed habitats in lakes. The radula formula is 19-1-19 to 21-1-21, and the cusps are the same as in *F. parallela*.

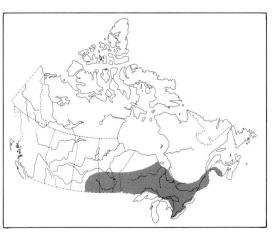

90
Ferrissia rivularis
a,b: Carp R. near Carp, Ont.; 4.8 mm.
c,d: Whitesand R. near Sheho, Sask.; 2.8 mm.

91 *Margaritifera margaritifera*

92 *Margaritifera falcata*

93 *Gonidea angulata*

94 *Amblema plicata*

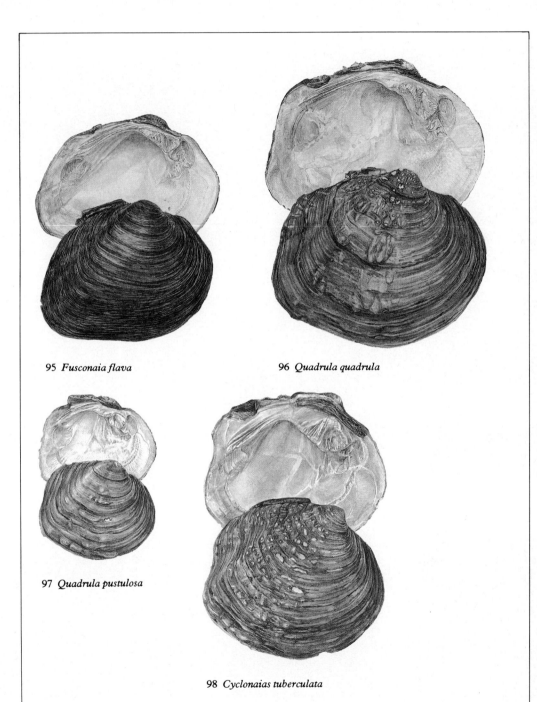

95 *Fusconaia flava*

96 *Quadrula quadrula*

97 *Quadrula pustulosa*

98 *Cyclonaias tuberculata*

99 *Elliptio complanata*

100 *Elliptio dilatata*

101 *Pleurobema coccineum*

102 *Alasmidonta viridis*

103 *Alasmidonta heterodon*

104 *Alasmidonta marginata*

105 *Alasmidonta undulata*

106 *Alasmidonta varicosa*

107 *Lasmigona complanata*

108 *Lasmigona compressa*

109 *Lasmigona costata*

110 *Simpsoniconcha ambigua*

111 *Anodontoides ferussacianus*

112 *Anodonta beringiana*

113 *Anodonta cataracta cataracta*

114 *Anodonta cataracta fragilis*

115 *Anodonta grandis grandis*

116 *Anodonta grandis simpsoniana*

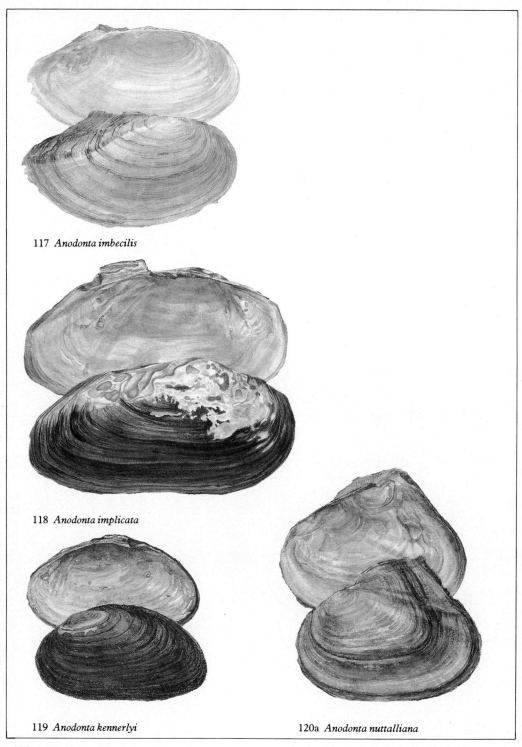

117 *Anodonta imbecilis*

118 *Anodonta implicata*

119 *Anodonta kennerlyi*

120a *Anodonta nuttalliana*

120b *Anodonta nuttalliana*

121 *Strophitus undulatus*

122 *Ptychobranchus fasciolaris*

123 *Obliquaria reflexa*

124 *Truncilla donaciformis*

125 *Truncilla truncata*

126 *Proptera alata*

127 *Carunculina parva*

128 *Obovaria olivaria*

129   *Obovaria subrotunda*

130   *Leptodea fragilis*

131   *Actinonaias carinata*

133 *Ligumia recta* ♀

132 *Ligumia nasuta* ♀

134 *Lampsilis cariosa* ♀

135 *Lampsilis fasciola*

136 *Lampsilis ochracea*

137 *Lampsilis radiata radiata*

138 *Lampsilis radiata siliquoidea*

139   *Lampsilis ventricosa* ♀

140   *Villosa fabalis*

141   *Villosa iris*

142   *Dysnomia torulosa rangiana* ♀

143   *Dysnomia triquetra* ♀

Class Pelecypoda
(Clams and Mussels)

Order Eulamellibranchia

IX Superfamily Unionacea
Freshwater Mussels

FAMILY MARGARITIFERIDAE
(Pearly River-Mussels)

Shell large, bivalved, of medium thickness, with pearly nacre, pseudocardinal hinge teeth well developed, and lateral hinge teeth only partly developed or absent. All 4 demibranchs (gills) are marsupial (have brood chambers for glochidia), but the marsupial microstructure is less complex and presumably more primitive than in Unionidae. The glochidia are hookless but have irregular small teeth at the ventral margin of the valves. The mantle is not united posteriorly and does not form separate siphonal openings. The family occurs only in North America and Eurasia.

91
Margaritifera margaritifera (Linnaeus, 1758)
Eastern-River Pearl Mussel

DESCRIPTION

Shell up to 150 mm long, 65 mm high, 40 mm wide, and with shell wall 10 mm thick at mid-anterior; long-oval, and with straight or concave lower margin. Except for closely spaced concentric lines the surface is mainly smooth. Periostracum brown in juveniles, black in adults, and eroded at umbones. Nacre white or whitish, with or without pink or purple suffusions. Beak sculpture consists of a few coarse ridges parallel to the lines of growth. Hinge teeth incomplete: pseudocardinals erect and serrated, 1 in the right valve and 2 in the left; lateral teeth poorly developed or absent. Muscle scars in beak cavity only partly visible from below. The sexes are separate.

The long and slightly bent shape, blackish periostracum, strong pseudocardinal teeth, and obscure lateral teeth serve to readily distinguish this species.

DISTRIBUTION

Atlantic drainage of North America from Goose Bay, Labrador, to the Little Schuylkill River, Pennsylvania. Also occurs in Eurasia from northern Spain and Scandinavia east through northern USSR to Japan. Especially abundant in Newfoundland and Nova Scotia, but does not virtually pave the bottoms of streams as the western species *M. falcata* sometimes does.

ECOLOGY

Occurs in small and medium-sized running streams. Often found on sandy shoals and in pools under overhanging branches. The breeding season occurs between June and August. The glochidia are oval to almost circular, with a narrow inward-directed flange at the ventral edge of each valve, and very small (0.06 mm long and 0.07 mm high). Host fishes in Europe are the brown trout and the minnow *Phoxinus phoxinus*. In North America the native brook trout, as well as the introduced brown trout, may be a host. In Europe the species reaches at least 116 years of age. In North America it appears to be restricted to soft water.

91
Margaritifera margaritifera
a,b,c,d: Burnt Berry Brook near Springdale, Nfld.; *a* and *d* 114.3 mm.

The specimen illustrated in colour on page 229 is from the Parrsboro R. near Parrsboro, N.S. (× 2/3).

92
Margaritifera falcata
(Gould, 1850)
Western-River Pearl Mussel

DESCRIPTION
Similar to *Margaritifera margaritifera* except smaller (maximum length 125 mm, with purple rather than white or whitish nacre, relatively smaller anterior pseudocardinal tooth in left valve, and muscle scars in beak cavity entirely visible from below. Hermaphroditic, whereas in *M. margaritifera* the sexes are separate.

DISTRIBUTION
Distributed in Pacific drainages from California and New Mexico north to the southern interior of British Columbia, and farther north near the coast to Revillagigedo Island in southern Alaska. Also in the upper Missouri drainage in Montana. In favourable localities in British Columbia the mussels may be so abundant and closely packed that they completely obscure the stream bed.

ECOLOGY
Occurs in running streams wider than 4 m. Found in sand or gravel substrates. Unlike *M. margaritifera*, it occurs in hard as well as in soft water. The gravid period is from mid May to late June. Hermaphroditic. The glochidia have not been described. Host fishes are the chinook salmon, rainbow trout, brown trout, brook trout, speckled dace, Lahontan redside, and Tahoe sucker. The greatest longevity so far determined for *M. falcata* is about 67 years, but older specimens probably occur.

92
Margaritifera falcata
a,b,c,d: Koksilah R. near Duncan, Vancouver Is., B.C.;
  a and *d* 101.6 mm.

The specimen illustrated in colour on page 229 is from the same locality (× 2/3).

FAMILY UNIONIDAE (Pearly Mussels)

Shells small to large, bivalved, thin and fragile to thick and heavy, with pearly nacre, and with or without pseudocardinal hinge teeth and/or lateral hinge teeth. All four demibranchs (gills), or only one pair of demibranchs, are marsupial, and the microstructure is more complex than in Margaritiferidae. The glochidia are hookless or with hooks. The mantle is drawn together posteriorly by the diaphragm, and separate siphonal openings are present. The family is worldwide but occurs principally in Europe, Asia and North America.

For many years North American Unionidae have been grouped into three subfamilies: Ambleminae (or Elliptioninae), characterized principally by heavy hinge teeth, all four demibranchs being marsupial, hookless glochidia, and a short breeding season; Anodontinae, with hinge teeth incomplete or absent, whole outer demibranchs only used as marsupia, hooked glochidia, and short breeding seasons; and Lampsilinae, with well-developed hinge teeth, posterior part of outer demibranchs only used as marsupia, hookless glochidia, and long breeding seasons. That classification is now being revised by several workers (see e.g. Heard and Guckert 1970), but the matter is still unsettled. In this book the traditional subdivisions of Unionidae are therefore retained.

SUBFAMILY AMBLEMINAE
(Button Shells and Relatives)

93
Gonidea angulata (Lea, 1839)
Rocky Mountain Ridged Mussel

DESCRIPTION
Shell up to 125 mm long, 65 mm high, 40 mm wide, and with shell wall up to about 5 mm thick at mid-anterior; variable in form but typically rather thin, trapezoidal in shape, with posterior margin obliquely flattened and relatively broad, and with a sharp and prominent posterior ridge running from the umbo to the angular basal posterior margin of each valve. Shell with obscure radial sculpturing on the posterior slope and readily apparent growth rests. Periostracum yellowish brown to blackish brown, without rays, smooth on the disc, and roughened on the posterior slope. Nacre centrally white or salmon, but pale blue posteriorly and near the margin. Beak sculpture composed of about 8 rather coarse, concentric ridges that are straight in the centre and curved at both ends. Hinge teeth irregular and poorly developed; pseudocardinal teeth compressed, low, laterally expanded, 1 in the right valve and none or 1 in the left; lateral teeth absent.

DISTRIBUTION
Columbia River system in southern British Columbia (Okanagan and Kootenay rivers) and south in the Pacific drainage to southern California.

ECOLOGY
In Vaseux Lake near Oliver, British Columbia, large specimens occur in muddy sand at a depth of 0.6 to 0.9 m along the shoreward edge of a bed of *Potamogeton* (pond weed). Elsewhere occurs in rivers and lakes and on various substrates. Four specimens were collected on 6 August 1972 from Vaseux Lake, and none were gravid. Nothing has been published about its reproduction, glochidia, or fish host.

93
Gonidea angulata
a,b,c,d: Vaseux L. near Oliver, B.C.; *a* and *d* 123.8 mm.

The specimen illustrated in colour on page 230 is from the same locality ($\times$ 2/3).

94
Amblema plicata (Say, 1817)
Three-Ridge

DESCRIPTION

Shell to about 150 mm long, 105 mm high, 65 mm wide, and with mid-anterior shell wall 15 mm thick; ovate, more or less truncated posteriorly, and with heavy sculpturing. Shells of most specimens bear 3 or 4 heavy, rounded, diagonal, more or less parallel centrally located ridges that are directed toward the lower posterior edge. In some old specimens the ridges are lacking. Surface also bears a few short ridges on the posterior slope perpendicular to the lines of growth and irregular concentric wrinkles over the whole shell. Periostracum brown to blackish and without rays. Nacre white and iridescent posteriorly. Beaks elevated above hinge line, located close to the anterior end, moderately inflated, and deeply excavated. Beak sculpture consisting of about 5 concentric single-looped ridges; early ridges expanded anteriorly. Hinge teeth massive and very strong; pseudocardinals thick and heavy, subtriangular, erect, deeply serrated, 2 or 3 in the right valve and 2 in the left; laterals of moderate length, straight or slightly curved, 1 in the right valve and 2 in the left.

Has the heaviest and most ponderous shell of any mollusc in Canada. Easily recognized by its impressive size and thickness, and by its wide diagonal ridges.

DISTRIBUTION

Occurs in the Red River-Lake Winnipeg drainage area, in the Great Lakes-St. Lawrence system from Lake Michigan and Lake Huron to Lake Erie (in the lakes themselves and in their tributaries), and in the whole Ohio-Mississippi River system.

ECOLOGY

Common in southwestern Ontario and in southern Manitoba. Typically a river species. Lives on or in various substrates. Stunted individuals occur in Lake Erie. A short-term breeder, with gravid period extending from May to July. Glochidia are oval, without hooks, and measure about 0.21 mm in length and 0.23 mm in height. Numerous fish hosts are known.

94
Amblema plicata
a,d: Grand R. near Dunnville, Ont.; 123.8 mm.
b,c: South Bay, Pelee Is., L. Erie, Ont.

The specimen illustrated in colour on page 230 is from McGregor Creek, Chatham, Ont. (× 2/3).

95
Fusconaia flava
(Rafinesque, 1820)
Pig-Toe

DESCRIPTION
Shell up to 100 mm long, 70 mm high, 40 mm wide, and with shell wall 8 mm thick at mid-anterior; roughly triangular, bluntly pointed at posterior basal region, with ventral margin straight or slightly concave posteriorly, and posterior ridge well marked. Periostracum with a dull lustre, brown in immature specimens and brownish black in adults, and without rays or with obscure rays only on the posterior slope. Nacre whitish or tinged with salmon. Umbones anterior of centre but in most specimens not close to anterior margin. Beak sculpture moderately fine and consisting of 3 to 5 small more or less concentric bars visible principally on the posterior ridge. Pseudocardinal hinge teeth moderately heavy and with radial furrow, 1 tooth in the right valve and 2 in the left; lateral teeth strong, of medium length, straight or slightly curved, 1 or 2 in the right valve and 2 in the left.

Sometimes confused with the less common species *Pleurobema coccineum*, with which it should be compared.

DISTRIBUTION
In Canada occurs only in the Lake Huron, Lake St. Clair and Lake Erie drainage basins of Ontario and in the Red River–Nelson River system of Manitoba. In the United States it lives in the same drainage systems but is also widespread in the Mississippi–Missouri system.

ECOLOGY
Relatively common within its range and always found in company with several other species of mussels. Occurs principally in medium or large rivers but also in large lakes, for example Lake Erie. Found in mud or sand and among sparse or moderate vegetation. Glochidia are ovate, hookless, and measure about 0.15 mm in height and width. Females are gravid in spring and summer, all four demibranchs are marsupial, and the host fish are bluegill, white crappie, and black crappie.

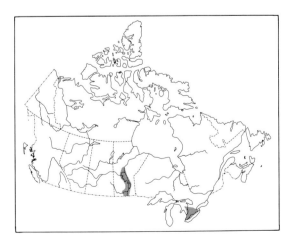

95
Fusconaia flava
a,d: Sydenham R. near Strathroy, Ont.; 92.1 mm.
b,c: Rondeau Harbour, L. Erie, Ont.

The specimen illustrated in colour on page 231 is from Big Creek near Port Rowan, Ont. (× 2/3).

96
Quadrula quadrula
(Rafinesque, 1820)
Maple-Leaf

DESCRIPTION
Shell up to 125 mm long, 100 mm high, 50 mm wide, and with shell wall 8 mm thick at the mid-anterior; more or less quadrate, rounded anteriorly, somewhat truncated posteriorly, and sculptured on each valve with 2 bands of nodules radiating from the umbones, 1 centrally located and the other on the posterior ridge. Periostracum yellowish brown or brown, and with poorly defined green radial bands in some specimens. Annual growth rests well defined. Nacre white. Beak sculpture consisting of tiny nodules. Hinge teeth strong: pseudocardinal teeth erect, conical, and serrated; 1 major tooth (with a small one in front and another behind) in the right valve and 2 in the left; lateral teeth erect, 1 or 2 in the right valve and 2 in the left.

Similar to *Q. pustulosa* except that the nodules are restricted to 2 radial bands, while in *Q. pustulosa* they occur over most of the shell surface. Also more quadrate and less ovate than *Q. pustulosa*.

DISTRIBUTION
Southern Ontario in the Lake Erie and Lake St. Clair drainages and Manitoba in the Red River system. In the United States it occurs in those drainage systems as well as throughout the Ohio-Mississippi drainage system.

ECOLOGY
Occurs in medium-to-large rivers where currents are slow to moderate. Usual substrates are mud or sand, and vegetation is ordinarily present and sometimes dense. Normally a minor element of unionid communities that contain several species. Adult females are gravid in late spring and summer, and the glochidia are ovate, without hooks, and measure about 0.08 mm in length and height. The host fish is the flathead catfish.

96
Quadrula quadrula
a,b,c,d: Grand R. near Byng, Ont.; *a* and *d* 84.1 mm.

The specimen illustrated in colour on page 231 is from the Grand R. near Dunnville, Ont. (× 2/3).

97
Quadrula pustulosa (Lea, 1831)
Warty-Back

DESCRIPTION
Shell up to 100 mm long, 85 mm high, 65 mm wide, and with shell wall 8 mm thick at the mid-anterior; ovate-quadrate, rounded in front, truncated behind, and sculptured with elongate tubercles that are parallel to the lines of growth and distributed chiefly over the central part of each valve. If tubercles are numerous some are also present on the posterior slope. Periostracum yellowish brown to chestnut-brown in adults and yellowish with greenish rays in young individuals. Annual growth rests well marked in Canadian specimens. Nacre white. Umbones prominent and, when not corroded, sculptured with 3 or 4 coarse ridges. Hinge teeth heavy: pseudocardinal teeth erect, triangular, and serrated, 1 large tooth in the right valve (with a small tooth before and behind) and 2 in the left; lateral teeth short and slightly curved, 1 or 2 in the right valve and 2 in the left.

Similar to *Q. quadrula* but is more ovate, has scattered rather than confined nodules, and is relatively heavier.

DISTRIBUTION
Lake Erie and Lake St. Clair drainages in southwestern Ontario and the United States. Also in the Lake Michigan and Ohio-Mississippi River drainages.

ECOLOGY
Not common in southwestern Ontario. Occurs in rivers of various widths. Found on gravel, sand, or mud bottoms. Females are gravid from mid May to late August. The glochidia are purse-shaped, without spines, and measure about 0.23 mm in length and 0.30 mm in height. The host fish are the channel catfish, shovelnose sturgeon, black bullhead, brown bullhead, flathead catfish, and white crappie.

97
Quadrula pustulosa
a,d: Grand R., Port Maitland, Ont.; 66.7 mm.
b,c: South Bay, Pelee Is., L. Erie, Ont.

The specimen illustrated in colour on page 231 is from L. Erie off Pelee Is., Ont. (× 2/3).

98
Cyclonaias tuberculata
(Rafinesque, 1820)
Purple Pimple-Back

DESCRIPTION

Shell about 100 mm long, 100 mm high, 50 mm wide, and with shell wall 8 mm thick anterio-centrally; outline circular except flattened dorsally and anteriorly, compressed, and with small or medium-sized nodules scattered over the posterior and central surfaces or more restricted. Nodules rounded or elongate, and oriented perpendicular to the lines of growth. Periostracum brown or blackish, with poorly defined brownish or greenish rays, and with annual growth rests well marked. Nacre purple. Beak sculpture of fine zigzag bars. Hinge teeth heavy and strong: pseudocardinal teeth massive and serrate, 1 in the right valve (with an additional small tooth in front and another behind) and 2 in the left; lateral teeth of medium length and nearly straight, 1 in the right valve and 2 in the left. Beak cavity narrow and deeply excavated.

The circular and compressed shape, nodulous exterior, massive pseudocardinal teeth, and purple nacre readily distinguish it from all other species.

DISTRIBUTION

Lake Erie and the Sydenham River in southern Ontario. In the United States it occurs in Lake St. Clair, Lake Erie, Lake Michigan and their drainage areas, and in the Ohio-Mississippi River drainage system.

ECOLOGY

An uncommon species in Canada. Occurs in rivers of various sizes. Found on gravel or mud bottoms. Gravid females have been collected in the United States from late May to the middle of August. The glochidia are reported as 0.27 mm long and 0.35 mm high, with short hinge line and rounded anterior, and without spines. The host fish is unknown.

98
Cyclonaias tuberculata
a,d: Huron R., Ann Arbor, Mich.; 79.4 mm.
b,c: South Bay, Pelee Is., L. Erie, Ont.

The specimen illustrated in colour on page 231 is from the Huron R., Mich. (× 2/3).

99
Elliptio complanata
(Lightfoot, 1786)
Eastern Elliptio

DESCRIPTION
Shell up to about 125 mm long, 65 mm high, 40 mm wide, and with shell wall 6 mm thick at mid-anterior; unusually variable in form but typically of medium thickness, somewhat trapezoidal or elliptical, obliquely truncated posteriorly, with rounded posterior ridge, and compressed. Shell unsculptured except for beak sculpture and lines of growth. Periostracum brownish or blackish and unrayed except in some young specimens and in adults from sandy substrates. Annual growth rests well marked. Nacre purple in most specimens, but pinkish or whitish in some. Beak sculpture of concentric U-shaped ridges. Hinge teeth well developed: pseudocardinal teeth compressed and conical, 1 in the right valve (a small accessory tooth may be present in front of the large one) and 2 in the left; lateral teeth narrow, of medium length and nearly straight, 1 in the right valve and 2 in the left. Muscle scars not deeply impressed.

Usually distinguished from *E. dilatata* by its compressed and trapezoidal shape, obliquely flattened posterior margin, beaks that are not close to the anterior end, and often by its thinner shell. More inflated specimens are broadest near the posterior slope, whereas *E. dilatata* is broadest in the anterior region, more regularly tapered posteriorly, and has beaks close to the anterior end. *Ligumia nasuta*, which also has purple nacre, differs from *E. complanata* in its narrower shape, more delicate hinge teeth, and centrally pointed posterior margin.

DISTRIBUTION
Southern James Bay drainages and the St. Lawrence system (except Lake Huron south of Georgian Bay, Lake Michigan, and most of Lake Erie) south in the Atlantic drainage to Georgia.

ECOLOGY
Uncommon in James Bay drainages but abundant elsewhere. Lives in shallow water of permanent lakes, rivers and medium-sized streams. Found on gravel, sand, clay, or mud bottoms. Females are gravid in late spring and early summer, and mature glochidia are about 0.20 mm long, subovate, and without hooks. The yellow perch is its only known host.

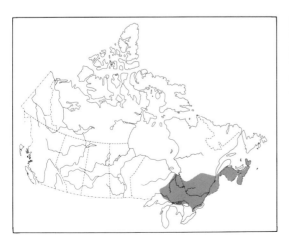

99
Elliptio complanata
a,d: Ottawa R. near Pembroke, Ont.; 77.8 mm.
b,c: Poucette L. near Port Elgin, N.B.

The specimen illustrated in colour on page 232 is from the Ottawa R. near Pembroke, Ont. (× 2/3).

100
Elliptio dilatata
(Rafinesque, 1820)
Spike, or Lady-Finger

DESCRIPTION

Shell up to 125 mm long, 65 mm high, 45 mm wide, and with shell wall up to 12 mm thick at mid-anterior; variable in form but normally long-ovate, rather thick and strong, and somewhat inflated. Shell unsculptured except for beak sculpture and lines of growth. In some specimens the growth lines are not evenly concentric, but are disproportionately wide apart posteriorly and give the shell a deformed appearance. Periostracum yellowish brown to brown, with indistinct greenish rays in many specimens, and in Canadian specimens with dark growth rests. Nacre purple in most specimens but white or pale pink in some. Beak sculpture consisting of 4 or 5 rather heavy curved bars. Hinge teeth thick and strong: pseudocardinal teeth conical and serrated, 1 in the right valve (with 1 small tooth in front and another behind) and 2 in the left; lateral teeth of medium length and nearly straight, 1 (or 2) in the right valve and 2 in the left. Muscle scars rather deeply impressed.

Sometimes confused with *E. complanata*. Compare with that species.

DISTRIBUTION

Common in the Great Lakes and their tributaries from Lake Michigan to Lake Erie; uncommon in Lake Ontario and in the St. Lawrence River. Widespread in the Ohio-Mississippi and Gulf of Mexico drainages.

ECOLOGY

Often abundant. Occurs in rivers and lakes of various sizes. Found on diverse substrates. Hermaphroditic. The breeding season lasts from May to August. Glochidia are oval except for the straight hinge line, and measure about 0.20 mm in length and 0.22 mm in height. The host fish are gizzard shad, flathead catfish, white crappie, black crappie, and yellow perch.

100
Elliptio dilatata
a,d: Grand R., Cayuga, Ont.; 85.7 mm.
b,c: South Bay, Pelee Is., L. Erie, Ont.

The specimen illustrated in colour on page 232 is from the Grand R. near West Montrose, Waterloo Co., Ont. ($\times$ 2/3).

101
Pleurobema coccineum
(Conrad, 1836)
False Pig-Toe

DESCRIPTION

Shell up to 90 mm long, 65 mm high, 40 mm wide, and with shell wall 8 mm thick at mid-anterior; more or less ovate, more sharply rounded posteriorly than anteriorly but not pointed, and with ventral margin convex throughout; posterior ridge obscure and rounded on most specimens. Periostracum shiny, yellowish brown to brown, but blackish in very old specimens, and without rays (or with obscure rays) only on the posterior slope. Nacre white or pinkish. Umbones in most specimens close to anterior margin. Beak sculpture coarse, irregular, and visible principally on posterior ridge. Pseudocardinal hinge teeth heavy and radially channelled and grooved, 1 large tooth in the right valve (with 1 additional small tooth in front and another behind) and 2 in the left; lateral teeth strong and straight or slightly curved, 1 in the right valve and 2 in the left.

Similar to the more abundant species *Fusconaia flava*, with which it should be compared.

DISTRIBUTION

Within Canada found only in Lake Erie and tributaries of Lake Erie and Lake St. Clair in southwestern Ontario. Farther south it occurs throughout most of the Mississippi-Missouri system.

ECOLOGY

Rare in Canada. Occurs in medium-sized to large rivers and in Lake Erie. Found on mud or sand, and with associated vegetation sparse or absent. Moderate current is apparent in its river localities. Gravid in late spring and early summer. The glochidia are subovate, without hooks, and measure about 0.15 mm in height and width. The outer demibranchs only are marsupial. The host fish is unknown.

101
Pleurobema coccineum
a,d: South Bay, Pelee Is., L. Erie, Ont.; 63.5 mm.
b,c: Rondeau Harbour, L. Erie, Ont.

The specimen illustrated in colour on page 232 is from Rondeau Harbour, L. Erie, near Shrewsbury, Ont. (× 2/3).

SUBFAMILY ANODONTINAE
(Floater Mussels)

102
Alasmidonta viridis
(Rafinesque, 1820)
Brook Wedge Mussel

DESCRIPTION
Shell up to 50 mm long (usually less), 30 mm high, 18 mm wide, and with shell wall up to 2 mm thick; more or less rhomboid in shape, quite thin, and somewhat inflated, especially at the rounded posterior ridge. Shell smooth except for lines of growth and beak sculpturing. Periostracum dull yellowish brown to greenish, and in many specimens covered with rather obscure greenish rays. Nacre white or bluish white, and iridescent posteriorly. Beak sculpture consists of 6 to 8 concentric ridges that are gently and unevenly curved centrally and angular on the posterior slope. Hinge teeth moderately small; pseudocardinal teeth elevated, triangular, 1 in the right valve and 1 split tooth in the left (a smaller tooth may also be present anterior to the major tooth); lateral teeth irregularly developed, 1 or 2 in the right valve and 2 in the left, but these may be reduced or nearly absent.

Alasmidonta viridis was formerly known as *A. calceola* (Lea, 1830).

DISTRIBUTION
Canadian specimens have come only from southern Ontario in the Lake Huron, Lake St. Clair, and Lake Erie drainages. In the United States, *A. viridis* occurs in the middle Great Lakes drainages and in the Mississippi system from the Ohio River drainage to the Tennessee River drainage.

ECOLOGY
Characteristic of small headwater streams, but also occurs in larger rivers and in lakes. Substrates are usually sand or gravel but sometimes mud. A long-term (bradytictic) breeder but details are not known. Glochidia are subtriangular, with a ventral spine on each valve, and measure about 0.30 mm long and 0.26 mm high. Host fish are the johnny darter and mottled sculpin.

102
Alasmidonta viridis
a,d: Irvine Creek near Dracon, Ont.
b,c: West branch of Grand R., Ayr, Ont.; 35.5 mm.

The specimen illustrated in colour on page 232 is from Smith Creek, Poole, Ont. (× 2/3).

103
Alasmidonta heterodon
(Lea, 1830)
Dwarf Wedge Mussel

DESCRIPTION

Shell up to about 45 mm long, 25 mm high, 16 mm wide, and with shell wall about 1 mm thick in mid-anterior region; more or less ovate or trapezoidal, roundly pointed posterio-basally, thin but not unduly fragile, with rounded posterior ridge, and of medium inflation. Females more inflated posteriorly than males. Sculpturing absent except for lines of growth and beak sculpture. Periostracum brown or yellowish brown, and with greenish rays in young or pale-coloured specimens. Nacre bluish or silvery white, and iridescent posteriorly. Beak sculpture composed of about 4 curved ridges, which are angular on the posterior slope. Hinge teeth small but distinct; pseudocardinal teeth compressed, 1 or 2 in the right valve and 2 in the left; lateral teeth gently curved and *reversed*, that is, in most specimens, 2 in the right valve and 1 in the left.

The small size, roundly pointed posterior-basal margin, and reversed lateral hinge teeth readily distinguish this species.

DISTRIBUTION

Discontinuously distributed in the Atlantic drainage from the Petitcodiac River in New Brunswick (its only Canadian population) to the Neuse River in North Carolina. Common in the Petitcodiac River and in portions of the Connecticut River system, but uncommon to rare elsewhere.

ECOLOGY

Characteristic of medium-sized rivers with slow-to-moderate current. Found on mud, sand, or (rarely) gravel bottoms. A long-term breeder, with gravid specimens recorded in February and April. Glochidia are roughly triangular, with hooks, and measure about 0.30 mm in length and 0.25 mm in height. The host fish has not been determined.

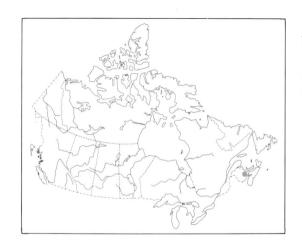

103
Alasmidonta heterodon
a,b,c,d: Ashuelot R. near Keene, N.H.; *b* and *c* 39.9 mm.

The specimen illustrated in colour on page 233 is from the North R. near Salisbury, N.B. (× 2/3).

104
Alasmidonta marginata
Say, 1819
Ridged Wedge-Mussel

DESCRIPTION
Shell up to about 90 mm long, 50 mm high, 35 mm wide, and with shell wall about 2.5 mm thick in mid-anterior region; roughly trapezoidal in shape, rather thin; and with a prominent, inflated, sharply rounded posterior ridge, and concave posterior slope. Shell sculptured on the posterior slope, with well-marked grooves and ridges perpendicular to the lines of growth. Periostracum yellowish, greenish, brownish or blackish, paler on the posterior slope, and prominently rayed in most specimens. Nacre bluish white, and exhibiting in some specimens greyish or greenish discolourations or suffusions of pale salmon. Beak sculpture coarse and consisting of 5 or 6 heavy double-looped bars. Hinge teeth characteristic: pseudocardinal teeth dorsoventrally compressed, well developed, 1 in each valve; lamellate interdental projection in left valve; lateral teeth absent.

The inflated posterior ridge, unique hinge teeth, sculptured posterior slope, and more or less trapezoidal shape distinguish it from all other species except *A. varicosa* of the Atlantic coastal region. In *A. varicosa* the posterior ridge is evenly (not sharply) rounded, the periostracum of some specimens is paler anteriorly than posteriorly (not the reverse), the shell does not exceed 75 mm in length, and it is not sharply truncated posteriorly.

DISTRIBUTION
Great Lakes-St. Lawrence system from Lake Huron to the Ottawa River and the vicinity of Cornwall, Ontario; Ohio-Mississippi River system and Susquehanna River system in the United States.

ECOLOGY
Occurs in rivers. Most common in riffles or rapids on gravel or rocky bottoms. Hermaphroditic. Gravid specimens have been noted in July; probably a long-term breeder. The glochidia bear hooks, and are about 0.34 mm in length and 0.37 mm in height. The host fish are white sucker, northern hog sucker, shorthead redhorse, rock bass, and warmouth.

104
Alasmidonta marginata
a,d: Nottawasaga R. near Alliston, Ont.; 78.6 mm.
b,c: Thames R., Chatham, Ont.

The specimen illustrated in colour on page 233 is from the Sydenham R. near Shetland, Ont. (× 2/3).

105
Alasmidonta undulata
(Say, 1817)
Heavy-toothed Wedge Mussel

DESCRIPTION
Shell up to 75 mm long, 45 mm high, 35 mm wide, and with shell wall about 6 mm thick at mid-anterior; triangular-ovate, centrally inflated, and with thickened anterior and low posterior ridge. Shell smooth except for lines of growth and heavy beak sculpture. Periostracum yellowish, greenish, reddish brown or black, and with greenish or blackish rays that are obscured in old blackened specimens. Nacre whitish anteriorly and bluish posteriorly, or modified with salmon or pink. Beak sculpture very heavy and composed of about 5 prominent single-looped curved ridges that extend far out on the disc (about 10 mm from the umbonal apex). Hinge teeth incomplete: pseudocardinal teeth strong and deeply grooved, 1 in the right valve and 2 in the left (the posterior one larger); interdental projection in left valve clearly apparent in many specimens; lateral teeth vestigial or absent. The pseudocardinal teeth are buttressed below by a heavy ridge located behind the impressed anterior adductor muscle scar.

The triangular form, modest size, thickened anterior, characteristic hinge teeth, and heavy beak sculpture distinguish this species from all others.

DISTRIBUTION
Atlantic drainage from Nova Scotia and the St. Lawrence River and its tributary rivers south to Florida.

ECOLOGY
Occurs in rivers and lakes. Found especially on sand or gravel bottoms. Reaches maximum size in outlet streams just below lakes. Breeds from the middle of July to the middle of the following June. Glochidia have strong hooks and measure about 0.34 mm in length and 0.36 mm in height. The fish host is unknown.

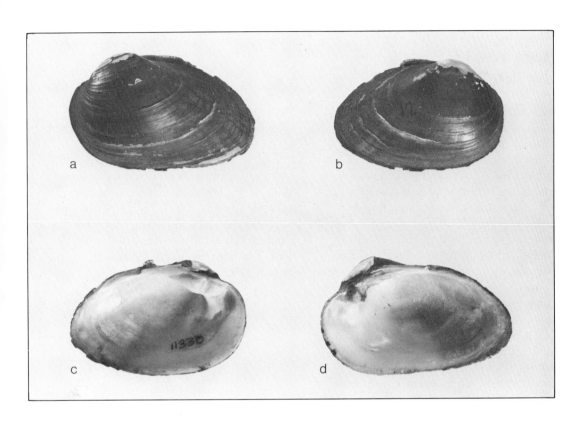

105
Alasmidonta undulata
a,b,c,d: St. Lawrence R. near Quebec, Que.; *a* and *d* 73.0 mm.

The specimen illustrated in colour on page 233 is from the St. Lawrence R. near Montreal, Que. (× 2/3).

106
Alasmidonta varicosa
(Lamarck, 1819)
Swollen Wedge-Mussel

DESCRIPTION

Shell up to about 70 mm long, 40 mm high, 30 mm wide and with shell wall about 2 mm thick in mid-anterior region; elliptical to trapezoidal but with flattened ventral margin and truncated dorso-posterior margin, rather thin, and with an inflated and rounded posterior ridge and slightly concave posterior slope. Shell smooth, or lightly sculptured on the upper posterior slope, with short grooves and ridges perpendicular to the lines of growth. Periostracum yellowish, greenish, brownish or blackish, and (in most specimens) extensively rayed. Nacre bluish white with olive or pinkish suffusions. Beak sculpture coarse and consisting of a few heavy single- or double-looped ridges, but these ridges are rarely preserved. Hinge teeth not well developed: pseudocardinal teeth weak, flattened, 1 in the right valve and 1 (smaller) in the left; interdental projection in left valve poorly developed or absent; lateral teeth absent.

Characters that distinguish it from the closely related *A. marginata* are discussed under that species. The inflated posterior ridge and complete absence of lateral teeth easily differentiate it from the other species that it partially resembles, that is *Lasmigona compressa* and *L. costata*.

DISTRIBUTION

Atlantic coastal drainage from Nova Scotia and New Brunswick to North Carolina.

ECOLOGY

Usually found in rapids or riffles on rocky or gravel substrates and in sandy shoals, that is habitats similar to those of *A. marginata*. More abundant in small rivers and creeks, whereas *A. marginata* is more abundant in larger streams. The breeding period lasts from August to the following May, and glochidia are similar to those of *A. marginata*. The host fish has not been determined.

106
Alasmidonta varicosa
a,b,c,d: Wallace R. near Pugwash, N.S.; *a* and *d* 57.2 mm.

The specimen illustrated in colour on page 233 is from Molunkus Stream, Macwahoc, Aroostook Co., Me. (× 2/3).

107
Lasmigona complanata
(Barnes, 1823)
White Heel-Splitter

DESCRIPTION

Shell up to 190 mm long, 125 mm high, 65 mm wide, and with shell wall 10 mm thick in mid-anterior; circular-trapezoid in juveniles but becoming ovate with age, diagonally truncate posteriorly, and thickened and rather strong anteriorly but thin and brittle posteriorly. A prominent and high dorsal projection (wing) occurs in juveniles and half-grown specimens. The wing is sculptured with radial ridges in some southern specimens but, except for growth rests and beak sculpture, the whole shell is unsculptured in Canadian material. Periostracum brown and indistinctly rayed in juveniles, blackish brown and unrayed in adults. Nacre white and tinged with bluish white posteriorly. Beak sculpture strong and composed of about 8 irregular or broken double-looped ridges that, in some specimens, are nodulous. Hinge teeth characteristic: pseudocardinal teeth large, thick, variable and irregular, 1 in the right valve (sometimes with 1 or 2 smaller teeth on either side of the large tooth) and 2 in the left; interdental projection present and low, or absent; lateral teeth vestigial or absent.

Easily recognized by its compressed ovate shell, prominent dorsal wing, heavy pseudocardinal teeth, and white nacre.

DISTRIBUTION

Lake Winnipeg-Nelson River system from Alberta to western Ontario; middle Great Lakes-St. Lawrence River system in tributaries of Lake Michigan, Lake St. Clair and Lake Erie; Ohio-Mississippi River system throughout; and Alabama River system.

ECOLOGY

Occurs in rivers of various widths greater than about 7.5 m. Always found on sandy or muddy bottoms. The breeding season lasts from at least August to May. Glochidia are subtriangular, with hooks, and measure from 0.28 to 0.34 mm in length and 0.30 to 0.34 mm in height. The host fish are carp, green sunfish, largemouth bass, and white crappie.

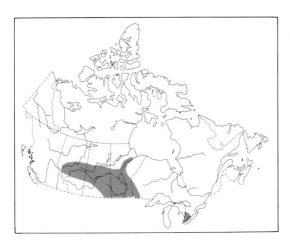

107
Lasmigona complanata
a,d: McGregor Creek, Chatham, Ont.; 127.0 mm.
b,c: Minnedosa R. near Minnedosa, Man.

The specimen illustrated in colour on page 234 is from the Seine R. near Winnipeg, Man. ($\times$ 2/3).

108
Lasmigona compressa
(Lea, 1829)
Brook Lasmigona

DESCRIPTION

Shell up to about 115 mm long, 55 mm high, 40 mm wide, and 4 mm in mid-anterior shell-wall thickness; trapezoid or rhomboid-ovate, with a low to moderate dorsal wing in immature specimens, compressed, and rather thin but not fragile. Rarely, a few faint radial grooves occur on the upper posterior slope, but ordinarily the shell is smooth except for beak sculpture and lines of growth. Periostracum yellowish brown, greenish or blackish brown, and extensively but not prominently rayed in most specimens. Nacre silvery white or bluish and, in some specimens, cream- or salmon-coloured near the beak cavities. Beak sculpture strong, and composed of about 8 variably irregular broken concentric ridges. Hinge teeth characteristic: pseudocardinal teeth strong, narrow, directed forward, typically 1 in the right valve and 2 in the left: interdental projection prominent in left valve; lateral teeth long and narrow, with 1 in the right valve and 2 in the left in most specimens, but undeveloped near the beaks or almost absent in some specimens.

Well characterized by its shape, colour, beak sculpture, hinge teeth, and especially its prominent interdental projection.

DISTRIBUTION

Hudson Bay drainage from Saskatchewan to Ontario; Great Lakes-St. Lawrence drainage from Minnesota to Quebec and Vermont; Hudson River system in New York; and upper Ohio-Mississippi system south to West Virginia and Nebraska.

ECOLOGY

Occurs principally in rivers and streams of various sizes, even sometimes in very small creeks, 2 m wide or less. Rare in lakes. Found on substrates of gravel, sand, or mud. The breeding season lasts from August to the following June. Normally hermaphroditic. The glochidia are circularly triangular, with hooks, and measure about 0.34 mm in length and 0.28 mm in height. The host fish is not known.

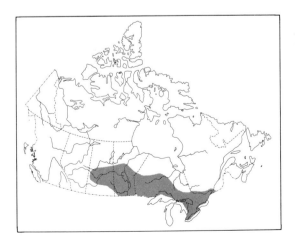

108
Lasmigona compressa
a,d: Creek near Leggatt, Ont.; 78.6 mm.
b,c: Turtlelake R., Edam, Sask.

The specimen illustrated in colour on page 234 is from the Red Deer R. near Hudson Bay, Sask. (× 2/3).

109
Lasmigona costata
(Rafinesque, 1820)
Fluted Shell

DESCRIPTION

Shell up to about 150 mm long, 75 mm high, 50 mm wide, and with shell wall 6 mm thick in mid-anterior region; trapezoid-ovate, without a posterior wing in all growth stages, of medium inflation, quite thick and strong, and heavily sculptured. The shell is sculptured on the posterior slope with up to 20 heavy radial ridges, with growth irregularities on the disc, and with beak sculpturing on the umbones. Periostracum yellowish, greenish or brownish, and with generally distributed narrow greenish or brownish rays that become mostly obscure in mature specimens. Nacre white or bluish white, with yellow or pink tints centrally. Beak sculpture composed of about 4 heavy double-looped concentric ridges. Hinge teeth well developed: pseudocardinal teeth strong and knob-like or lamellar, 1 in the right valve and 2 smaller ones in the left; interdental projection heavy and prominent in the left valve; lateral teeth rudimentary or absent.

The large size, characteristic shape, and heavy posterior sculpturing make identification of most specimens very easy. Small specimens from some populations with reduced sculpture may resemble *L. compressa*, but they can be distinguished by characters of the hinge teeth, beak sculpture, and internal structure (that is relative proportion of males to females).

DISTRIBUTION

Hudson Bay drainage in the Red and Winnipeg River systems; Great Lakes-St. Lawrence system from southern Lake Huron and its tributaries to the Ottawa River and Lake Champlain; and entire Ohio-Mississippi River system.

ECOLOGY

Occurs in canals, rivers, and lakes. Found on gravel, sand, or mud bottoms. Like other unionids it responds positively to increased water hardness by becoming unusually large and thick. The breeding season lasts from the beginning of August to the middle of May. The sexes are separate. Glochidia are triangular, and measure about 0.36 mm in length and 0.38 mm in height. A host fish is the carp.

109
Lasmigona costata
a,d: Grand R. near Doon, Ont.; 109.5 mm.
b,c: Baker Creek near Niagara R. near Fort Erie, Ont.

The specimen illustrated in colour on page 235 is from the St. Lawrence R. near Cornwall, Ont. (× 2/3).

110
Simpsoniconcha ambigua
(Say, 1825)
Mudpuppy Mussel

DESCRIPTION
Shell up to about 42 mm long, 20 mm high, 16 mm wide, and with shell wall almost 3 mm thick at mid-anterior; elongate elliptical, and sharply rounded at the ends but more or less straight dorsally and ventrally. Periostracum brownish and without rays. Nacre whitish or with muted yellow or purple suffusions, especially near the beak cavity. Beak sculpture of 4 or 5 parallel, V-shaped bars with their apices directed toward the umbonal apex. Beaks sharp, narrow, inclined forward, and located about 1/4 the distance from anterior to posterior. Hinge teeth incomplete and irregular: 1 pseudocardinal tooth in each valve; lateral teeth either rudimentary or lacking entirely.

This small, rare species resembles *Carunculina parva*, but may be distinguished by the incomplete hinge teeth and characteristic beak sculpturing. Compare also with *Alasmidonta heterodon* and *Villosa fabalis*.

DISTRIBUTION
Lake St. Clair drainage (Sydenham River, one record) in southern Ontario; central Great Lakes drainages in the United States; and Ohio-Mississippi drainage from Michigan to Iowa and south to Arkansas and Tennessee.

ECOLOGY
Usually found under flat stones in rivers but also in mud or on gravel. The glochidial host is an amphibian, the mudpuppy (*Necturus maculosus*). This is the only freshwater mussel known to utilize an animal other than a fish for glochidial attachment.

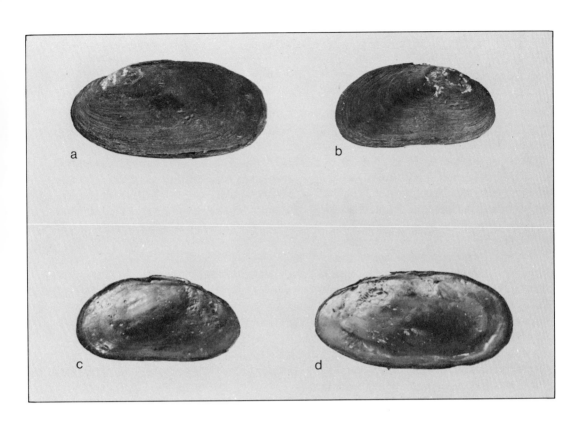

110
Simpsoniconcha ambigua
a,b,c,d: Troublesome R. above Lost Creek, Ky.; *a* and *d* 33.3 mm.

The specimen illustrated in colour on page 235 is from the same locality (× 2/3).

111
Anodontoides ferussacianus
(Lea, 1834)
Cylindrical Floater

DESCRIPTION

Shell up to 95 mm long, 40 mm high, 40 mm wide, and with shell wall 1.5 mm thick in mid-anterior region; elliptical, much inflated, rather fragile, and with unusual oblique beak sculpturing. Shell unsculptured except for concentric lines and low ridges associated with growth rests and beak sculpture. Periostracum greenish or brownish, with prominent dark annuli; numerous, generally distributed, prominent or obscure, narrow green or brown rays and 2 or 3 broader dark rays on the low rounded posterior ridge and posterior slope. Nacre bluish white, slightly iridescent, and with tints of cream in the beak cavity of some specimens. Beak sculpture unique, fine, and composed of several curved ridges that are not parallel to the lines of growth but follow an anteriorly expanded arc. Several fine radial ridges also exist posterior to the curved ridges. Hinge teeth absent except for a narrow swelling of the shell margin in front of the beak.

   Characterized by its moderate size; thin, inflated and subcylindrical shell; and especially by its fine, oblique, subconcentric beak sculpture.

DISTRIBUTION

James Bay and Hudson Bay drainage from central Ontario to southeastern Saskatchewan; Great Lakes-St. Lawrence system downstream to near Montreal; and Ohio-Mississippi River system south to Colorado and Tennessee.

ECOLOGY

Usually found in slow-moving streams on mud bottoms but also occurs in lakes and sometimes on sand. The breeding season lasts from August until May. Glochidia are subtriangular, with hooks, and measure about 0.32 mm in length and height. The mottled sculpin and the sea lamprey are host fish.

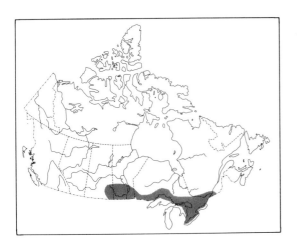

111
Anodontoides ferussacianus
a,d: Baker Creek near Niagara R. near Fort Erie, Ont.; 69.9 mm.
b,c: Grand R., Riverview, Ont.

The specimen illustrated in colour on page 235 is from the Souris R. near Souris, Man. (× 2/3).

112
Anodonta beringiana
Middendorff, 1851
Yukon Floater

DESCRIPTION
Shell up to 150 mm long, 75 mm high, 55 mm wide, and with shell wall about 3 mm thick at mid-anterior; elliptical, broadly rounded anteriorly and more narrowly rounded posteriorly, without a dorsal wing, and moderately thin but relatively strong. Surface roughened by lines of growth but centrally shining. Periostracum olive-green in juveniles but dark brown to nearly black in old specimens. Nacre lead-colour to dull blue. Umbones inflated and elevated above hinge line. Beak sculpture consisting of a few rather straight, irregular bars parallel to the hinge line. Hinge teeth absent.

Quite similar to very large specimens of *A. kennerlyi* but distinguished by its larger shell, dark periostracum, inflated beaks that clearly project above the hinge line, and lead-coloured to bluish nacre. Occurs in more northern drainages than *A. kennerlyi*.

DISTRIBUTION
Yukon River drainage in the Yukon Territory and Alaska, and other drainages in Alaska and in Kamchatka, USSR.

ECOLOGY
Known from rivers and lakes within its range. The glochidia have been reported as 0.296 mm in height and width. Fish hosts are the sockeye salmon, chinook salmon, and three-spined stickleback.

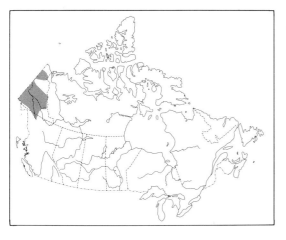

112
Anodonta beringiana
a,b: Old Crow R. near Old Crow, Y.T.; 133.4 mm.

The specimen illustrated in colour on page 236 is from a lake in Porcupine Valley near Fort Yukon, Y.T. (× 2/3).

113
Anodonta cataracta cataracta
Say, 1817
Eastern Floater

DESCRIPTION
Shell up to 150 mm long, 75 mm high, 65 mm wide, and with shell wall about 1.5 mm thick at mid-anterior; elliptical, roundly pointed posteriorly, inflated, thin, fragile, and with inflated beaks that project above the hinge line. Surface smooth except for low concentric wrinkles and growth rests. Periostracum shiny and grass-green, or modified with yellow or brown, or simply brown all over; in many specimens with numerous narrow green rays covering the disc and the central area, and a few broad green rays on the posterior ridge and slope. Nacre silvery or white, tinged with blue or yellow, and iridescent. Beak sculpture consisting of about 6 to 8 primarily double-looped, curved, concentric bars that are not nodulous. Hinge teeth absent.

Similar to *A. c. fragilis*, but that subspecies has beak sculpture with 8 to 12 single-looped bars and lacks green periostracum (which is present in many, but not all, *A. c. cataracta*). Compare also *A. grandis grandis* and *A. g. simpsoniana*.

DISTRIBUTION
Lower St. Lawrence River drainage and Maritime Provinces south in the Atlantic drainage to the Gulf of Mexico drainage. At the north end of its range it intergrades with *A. c. fragilis*, and in the middle of the St. Lawrence system with *A. grandis grandis*.

ECOLOGY
Occurs in ponds, lakes, and streams of various widths down to small brooks. Most abundant on mud (where it is commonly bright green) but also occurs on sand and, less frequently, on gravel. The breeding season lasts from July to the following April or May. Glochidia are roughly triangular, with hooks, and measure about 0.36 mm in length and 0.37 mm in height. The host fish are reported to be the pumpkinseed and the carp.

113
Anodonta cataracta cataracta
a,d: Emerald L., North Dorset, Vt.; 101.6 mm.
b,c: L. Edward, Denmark Parish, N.B.

The specimen illustrated in colour on page 236 is from a ditch near Quebec, Que. (× 2/3).

114
Anodonta cataracta fragilis Lamarck, 1819
Newfoundland Floater

DESCRIPTION

Shell up to about 90 mm long, 45 mm high, 25 mm wide, and with mid-anterior shell wall 1.5 mm thick; long elliptical with bluntly pointed posterior located at or below the midline, rounded or nearly straight ventrally, diagonally flattened dorso-posteriorly, and typically thin and fragile. Surface marked by fine concentric wrinkles and prominent growth rests. Periostracum shiny or dull, straw-yellow to brown, and (in a few pale specimens) with obscure fine green rays on the disc and central area and a few poorly defined broad green rays on the posterior ridge and slope. Nacre silvery, white or bluish, and centrally discoloured with yellowish blotches in most specimens. Beak sculpture consisting of about 8 to 12 fine, irregular, concentric, single-looped bars that extend about 8 mm from the umbonal apex. Hinge teeth absent.

Characterized principally by its beak sculpture, lack of green periostracum, and small size. Compare with *A. c. cataracta* and *A. kennerlyi*.

DISTRIBUTION

Typical specimens occur principally in Newfoundland, but specimens exhibiting intermediate characteristics between *A. c. fragilis* and *A. c. cataracta* are common throughout northern Nova Scotia, New Brunswick, and eastern Quebec. Such specimens are considered to be intergrades and are referred to simply as *Anodonta cataracta* or, more precisely, as *A. c. cataracta* × *A. c. fragilis*, or the reverse, with the dominant morph cited first.

ECOLOGY

Occurs in diverse, permanent-water habitats (ponds, lakes, and streams of various sizes) as docs *A. c. cataracta*. Found principally in mud but also in sand and, less commonly, in gravel. Undoubtedly a long-term breeder, but no details are known about its reproductive period, glochidia, or fish host.

114
Anodonta cataracta fragilis
a,d: Wells Gully, Whitbourne, Nfld.; 69.9 mm.
b,c: Poucette L. near Port Elgin, N.B.

The specimen illustrated in colour on page 236 is from the same locality as *a* and *d* ($\times$ 2/3).

115
Anodonta grandis grandis
Say, 1829
Common Floater

DESCRIPTION

Shell up to about 160 mm long, 100 mm high, 75 mm wide, and with shell wall 4 mm thick at mid-anterior and 8 mm thick near anterior pallial line. Shell highly variable in form but typically ovate, inflated, thin, and fragile. Surface smooth and shiny except roughened with low, concentric wrinkles and growth rests. Periostracum yellowish brown, greenish, greenish brown or blackish, and (in many specimens) with extensive but poorly defined green rays and concentric lighter and darker bands. Nacre white or bluish white, rarely pinkish. Umbones inflated, projecting well above hinge line, and located about 30% of the distance from anterior to posterior. Beak sculpture variable but ordinarily heavy, double-looped, and with the loop apices elevated and forming 2 radial rows of tubercules. Hinge teeth absent.

The above description is of typical *A. g. grandis*. Most specimens from the lower St. Lawrence system are much smaller (about 75 mm long), are relatively more elongated, have a dark and roughened periostracum, and exhibit beak sculpture in which the nodules are poorly developed. Typical *A. g. grandis*, however, is characterized by its ovate inflated form and its double-looped and nodulous beak sculpture.

DISTRIBUTION

Canadian Interior Basin from central Ontario to central Alberta; Great Lakes-St. Lawrence system east to near Montreal; Ohio-Mississippi system throughout; and Gulf of Mexico drainages in Louisiana and Texas.

ECOLOGY

Occurs in permanent ponds, lakes and rivers of various sizes. Found on all types of substrates but is most abundant on mud. The breeding season is reported to last from early August to the following April or May. Both dioecious (separate sexes) and monoecious (hermaphroditic) specimens occur. The glochidia are triangular-ovate, with spines, and measure from 0.31 to 0.36 mm long and 0.28 to 0.33 mm high. Numerous fish species have been shown to act as hosts (see Fuller *in* Hart and Fuller 1974).

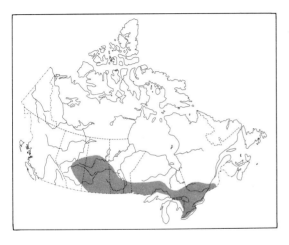

115
Anodonta grandis grandis
a,d: Gatineau R. near Kazabazua, Que.; 120.7 mm.
b,c: L. Erie, Point Pelee, Ont.

The specimen illustrated in colour on page 237 is from the Grand R. near Onondaga, Ont. (× 2/3).

116
Anodonta grandis simpsoniana Lea, 1861
Northern Floater

DESCRIPTION
Shell up to about 125 mm long, 55 mm high, 45 mm wide, with shell wall 2 mm thick at mid-anterior; variable but commonly elliptical, roundly pointed posteriorly, of medium inflation, thin, and rather fragile. Surface roughened by fine concentric wrinkles and prominent growth rests. Periostracum brown in most specimens, greenish or yellowish in others, and with faint greenish rays and/or concentric darker and paler bands in some specimens. Nacre silvery, white, or bluish, and with or without salmon or yellow suffusions near the beak cavity. Beaks low but clearly projecting above hinge line and located about 25% of the distance from anterior to posterior. Beak sculpture composed of 4 to 6 single-looped or faintly double-looped curved bars that are not nodulous. Hinge teeth absent.

May be distinguished from *A. g. grandis* principally by its beak sculpturing that is predominantly single-looped and not nodulous; in *grandis s. str.* it is double-looped *and* nodulous. Beak also farther forward and shell relatively more elongate and compressed than in *A. g. grandis*. Intergrades occur between the two subspecies in their zone of contact.

DISTRIBUTION
Canadian Interior Basin in the boreal forest region from northern Quebec west to central Alberta, and northwest to the mouth of the Mackenzie River.

ECOLOGY
Has been found in permanent ponds, in lakes, and in rivers more than about 9 m wide. All kinds of substrates are inhabited by this mussel. Gravid specimens with glochidia have been collected between July 22 and August 24, but the duration of its gravid period is not known. Glochidia are triangular-ovate, with hooks, and measure about 0.36 mm long and 0.35 mm high. The host fish is unknown.

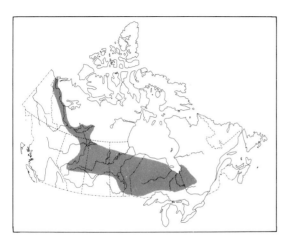

116
Anodonta grandis simpsoniana
a,d: L. St. Joseph near Rat Rapids, Ont.; 101.6 mm.
b,c: L. Caché near Chibougamau, Que. (49°50′N, 74°24′W).

The specimen illustrated in colour on page 237 is from Wilson L. south of Nellie Lake, Ont. (× 2/3).

117
Anodonta imbecilis Say, 1829
Paper Pond-Shell

DESCRIPTION

Shell up to about 90 mm long, 50 mm high, 40 mm wide, and with shell wall 1 mm thick at mid-anterior; elliptical, moderately inflated, pointed at mid-posterior, slightly alate, and fragile. Shell smooth except for very fine concentric striae, growth rests, and beak sculpture. Periostracum green to greenish brown, shiny, with numerous fine green rays visible on the disc, and with a few broad green or brown rays clearly apparent on the posterior ridge and slope. Nacre bluish white, silvery, and iridescent posteriorly. Umbones flat and without projections above the hinge line. Beak sculpture composed of about 6 fine, irregular, concentric ridges, with the earlier ones ridged and broken centrally, and the later ones somewhat double-looped. Hinge teeth absent.

The flat unelevated beaks, greenish fragile shell, and characteristic beak sculpturing distinguish this species from all other *Anodonta* in Canada and the northern United States.

DISTRIBUTION

Lake Erie and Lake St. Clair drainages in southern Ontario (Grand and Sydenham rivers) and adjacent parts of the United States; Ohio-Mississippi system generally; Gulf of Mexico drainage from the Ochlocknee River system in western Florida to the Rio Grande River system; and southern Atlantic coastal drainage from the Altamaha River system in Georgia to the Gunpowder River system in Maryland.

ECOLOGY

Occurs in muddy to somewhat sandy habitats in slow-moving rivers, canals, and lakes; rare on gravel bottoms. Hermaphroditic. A long-term breeder, with gravid periods probably overlapping in the summer in different individuals. Glochidia are roughly triangular, with hooks, and measure about 0.23 mm in length and height. The creek chub and the green sunfish have been implicated as host fish.

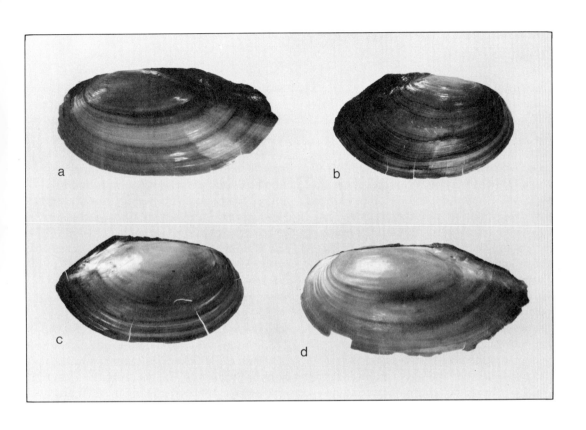

117
Anodonta imbecilis
a,b,c,d: Ohio Canal, Columbus, Ohio; *a* and *d* 69.9 mm.

The specimen illustrated in colour on page 238 is from the Erie Barge Canal near Macedon, N.Y. (× 2/3).

118
Anodonta implicata Say, 1829
Alewife Floater

DESCRIPTION

Canadian specimens are up to about 125 mm long, 65 mm high, 40 mm wide, and with shell wall about 3.2 mm thick at mid-anterior and near the anteroventral margin below the pallial line. (More-southern specimens may be 20% larger and have a shell wall 5 mm thick.) Shell long-elliptical, thickened anteriorly, thinner posteriorly, inflated in females and more compressed in males, and with the posterior ridge well defined and double in most specimens. Surface marked with prominent concentric growth rests, and in most specimens with severe corrosion on the anterior and around the umbones. Periostracum heavy, yellowish, brownish or blackish, and in some young specimens obscurely rayed. Nacre ordinarily salmon or pinkish or, more rarely, white or bluish. Beaks inflated and somewhat elevated. Beak sculpture composed of about 8 heavy double-looped concentric bars. Hinge teeth absent.

Adult is well characterized by its prominent anteroventral thickening below the pallial line, large size, dark, rayless periostracum, and salmon or pinkish nacre.

DISTRIBUTION

Atlantic coastal plain from Nova Scotia (Cape Breton) and eastern Quebec south to the Potomac River system in Maryland.

ECOLOGY

Restricted to coastal streams and lakes that can be reached by its anadromous host fish, the alewife. Occurs principally in sand and gravel, rarely in mud; the largest individuals have been found in relatively rapid streams. The breeding season is not known in detail but probably occurs from about August or September to June. Glochidia are rather large, approximately triangular, and with hooks at the tip of each valve. The host fish is the alewife, a predominantly saltwater fish that in the spring migrates into fresh water to spawn. Some true freshwater fish (white sucker, white perch, and pumpkinseed) have also been named as hosts, but the distribution of *A. implicata* makes this doubtful. Landlocked freshwater populations of the alewife also occur in the Great Lakes, but *A. implicata* is not found there.

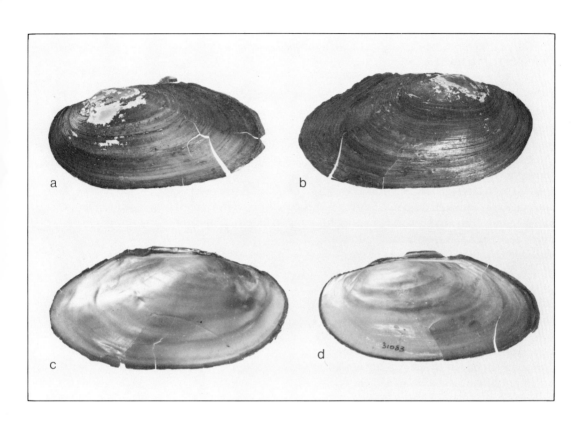

118
Anodonta implicata
a,d: Darlings L., Lakeside, N.B.
b,c: Branch of Denny Stream near Moores Mills, N.B.; 92.1 mm.

The specimen illustrated in colour on page 238 is from Great Herring Pond near Buzzard's Bay, Mass. (× 2/3).

119
Anodonta kennerlyi Lea, 1860
Western Floater

DESCRIPTION
Shell up to about 120 mm long, 65 mm high, 40 mm wide (but commonly much smaller), and with shell wall about 1.5 mm thick at mid-anterior and 3 mm at antero-ventral pallial line; elliptical, bluntly pointed posteriorly, without a dorsal wing, and relatively thin and fragile. Surface roughened by lines of growth but shiny in many specimens. Periostracum yellowish, yellowish brown or brown, tinged with green in some specimens, and with prominent dark-brown growth rests. Nacre whitish or bluish white, and in some individuals centrally suffused with salmon. Umbones flattened and barely projecting above hinge line. Beak sculpture consisting of about 15 irregular concentric ridges that extend up to 10 mm beyond the umbonal apex. Hinge teeth absent.

The elliptical shape and lack of a dorsal wing separate this species from the only other *Anodonta* in British Columbia, namely *A. nuttalliana*. Similar to *A. beringiana* but smaller and with lighter periostracum. Compare also with *A. cataracta fragilis*.

DISTRIBUTION
In British Columbia occurs abundantly on Vancouver Island and other coastal islands (including the Queen Charlotte Islands) and on the mainland from the Columbia to the Fraser and Skeena River systems. Also crosses the Continental Divide and occurs in a few mountain lakes in the uppermost North Saskatchewan and Athabasca River systems of Alberta. Southward the species extends in the Pacific drainage to Oregon.

ECOLOGY
Often abundant. Occurs in muddy or sandy substrates in rivers and lakes. Probably a long-term breeder, with the breeding period beginning in early August. The glochidia are triangular, with a straight hinge and a spine at the ventral apex of each valve, and measure approximately 0.30 mm in length and height. The fish host is unknown.

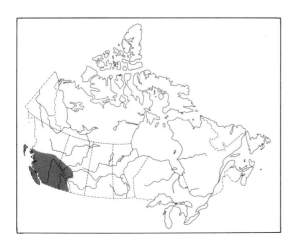

119
Anodonta kennerlyi
a,d: Aberdeen L., Aberdeen, Wash.; 93.7 mm.
b,c: Still Creek, Queen Charlotte Islands, B.C.

The specimen illustrated in colour on page 238 is from Horse L. east of 100 Mile House, Lillouet Dist., B.C. ($\times$ 2/3).

120
Anodonta nuttalliana Lea, 1839
Winged Floater

DESCRIPTION

Shell up to about 110 mm long, 75 mm high, 45 mm wide, and with mid-anterior shell wall about 3 mm thick; highly variable in shape but ordinarily trapezoid-ovate, centrally inflated, with posterior margin obliquely flattened, and with a more or less prominent dorsal wing. Some specimens are relatively compressed, and in others the dorsal wing is not well developed. Surface roughened by lines of growth. Periostracum yellowish green, yellowish brown or brown, and with prominent dark-brown growth rests in some populations. Nacre white or bluish. Umbones flattened and projecting only very slightly, or not projecting, above hinge line. Beak sculpture consisting of up to 20 or more fairly strong concentric ridges that may be irregularly single- or double-looped and that extend about 10 mm beyond the tip of the umbones. Hinge teeth absent.

The ovate and winged shape of *A. nuttalliana* easily distinguishes it from the other two Pacific-drainage *Anodonta* species in Canada, namely *A. kennerlyi* and *A. beringiana*, which are elliptical and lack a dorsal wing. The numerous umbonal ridges are characteristic of both *A. kennerlyi* and *A. nuttalliana* and differentiate them from all other Canadian species except *A. cataracta fragilis* from the Maritime provinces.

Synonyms include *A. wahlamatensis* Lea, 1839, and *A. oregonensis* Lea, 1839. In many localities *A. nuttalliana* occurs together with *A. kennerlyi* Lea, but intergrades have never been observed. The two species are therefore fully distinct.

DISTRIBUTION

Fraser and Columbia River systems in southern British Columbia, and south to central California. Absent from Vancouver Island.

ECOLOGY

Occurs in rivers and lakes on muddy and sandy bottoms. The largest Canadian specimens seen are from Vaseux Lake and Osoyoos Lake, both river-lakes in the Okanagan River, a tributary of the Columbia River. Little is known about its breeding season, although gravid specimens with immature larvae have been observed in October. The glochidia and host fish are unknown.

120
Anodonta nuttalliana
a,d: Osoyoos L., Osoyoos, B.C.; 114.3 mm.
b,c: Okanagan L. near Vernon, B.C.

The specimens illustrated in colour on pages 238 and 239 are from: a) Box L. south of Nakusp, B.C.;
b) Vaseux L. north of Oliver, B.C. (× 2/3).

121
Strophitus undulatus (Say, 1817)
Squaw-Foot

DESCRIPTION

Shell up to about 100 mm long, 55 mm high, 45 mm wide, and with mid-anterior shell wall about 3 mm thick; elliptical to trapezoid, somewhat compressed to moderately inflated, and of medium thickness. Surface roughened by concentric wrinkles and growth rests. Periostracum blackish brown in adults (paler near the umbones), and yellowish or greenish brown and with green rays in juveniles. Nacre white or bluish white, with suffusions of yellow or salmon near the beak cavity, and with a narrow band of olive-green or greenish brown at the border. Beak sculpture of 4 or 5 coarse concentric ridges that are approximately parallel to the growth lines. Hinge teeth rudimentary: pseudocardinal hinge teeth vestigial and indicated by a swelling of the hinge line on each valve just anterior to the beak or, rarely, by small clearly discernable pseudocardinal teeth; lateral teeth entirely absent.

The dark periostracum, vestigial pseudocardinal hinge teeth, and characteristic nacre readily distinguish this species. It is closest in appearance to *Anodontoides ferussacianus*, but that species is paler in colour, has a thinner shell, and has fine, clearly oblique beak sculpture.

DISTRIBUTION

Canadian Interior Basin in the Red River–Nelson River system from western Ontario to eastern Saskatchewan; Great Lakes–St. Lawrence system throughout; Ohio-Mississippi system from Minnesota to Texas and from Pennsylvania to Tennessee; and Atlantic coastal drainage from Nova Scotia to South Carolina.

ECOLOGY

Occurs principally in rivers and creeks but occasionally in lakes. Inhabits all substrates, but the finest specimens are from sandy bottoms, especially in lake outlets. The breeding season lasts from July to the following April or May. The glochidia are subtriangular, with hooks, and have been reported as 0.36 mm long and 0.30 mm high in one population, and as 0.46 mm long and 0.36 mm high in another. The glochidia are reputed to be able to complete their development without a period of parasitism on fish. Successful metamorphosis has also been reported *on* fish, however, in particular on the largemouth black bass and the northern creek chub.

121
Strophitus undulatus
a,d: Creek near Leggatt, Ont.; 63.5 mm.
b,c: L. Winnipeg near Elk Is., Man.

The specimen illustrated in colour on page 239 is from the Assiniboine R. near Amsterdam, Sask. ($\times$ 2/3).

SUBFAMILY LAMPSILINAE
(Lamp Mussels)

122
Ptychobranchus fasciolaris
(Rafinesque, 1820)
Kidney Shell

DESCRIPTION
Shell up to about 100 mm long, 65 mm high, 30 mm wide, and with mid-anterior shell wall about 6 mm thick; elliptical but straight ventrally, with ventral posterior end more or less drawn out in old specimens, compressed, thick, and solid. Surface unsculptured except for growth rests and roughened posterior slope. Periostracum yellowish, yellowish brown or brown, and with generally distributed broad, interrupted, greenish rays in many specimens. These rays are actually groups of very fine rays that, with the interruptions, appear as squarish spots. Nacre white or bluish white, but pinkish in some young specimens. Beak sculpture restricted to umbonal apices and consisting of poorly defined, broken short bars that appear as low double nodules. Hinge teeth thick and heavy: pseudocardinal teeth stumpy and medium-sized, 1 (or 2) in the right valve and 2 in the left; lateral teeth thick, inclined postero-ventrally, almost pendulous distally, 1 in the right valve and 2 in the left. The left lateral teeth converge near the umbones. The shells of females are marked internally by a prominent oblique groove that runs from the beak cavity diagonally across the middle of the shell toward the posteroventral end. This groove corresponds to the marsupium.

Usually distinguished by the interrupted green rays, elliptical and compressed shape, and heavy hinge teeth. Rayless specimens resemble *Elliptio dilatata*, but that species commonly has a purple nacre and lateral hinge teeth that are less massive and not distally pendulous.

DISTRIBUTION
In Canada known only in southern Ontario from western Lake Ontario, Lake Erie, and the tributaries of Lake St. Clair and Lake Erie. In the United States it occurs in these same drainages and throughout most of the Ohio-Mississippi system.

ECOLOGY
Uncommon in Canada. Usually found in river locations where currents are swift enough to maintain sand or gravel. Also occurs in lakes on sandy bottoms. Its breeding period lasts from early August to about the end of June. Glochidia are small, purse-shaped, without hooks, and higher than long. The host fish is unknown.

122
Ptychobranchus fasciolaris
a,d: Grand R., York, Ont.; 85.7 mm.
b,c: Thames R., Windsor, Ont.

The specimen illustrated in colour on page 239 is from the Sydenham R. near Shetland, Ont. (× 2/3).

123
Obliquaria reflexa
Rafinesque, 1820
Three-horned Warty-Back

DESCRIPTION

Canadian specimens are up to 53 mm long, 40 mm high, 30 mm wide, and with mid-anterior shell wall about 6.5 mm thick. (Specimens from more-southern localities may be up to nearly 75 mm long and with mid-anterior shell wall 10 mm thick.) Shell trapezoid-ovate, small, thick, and sculptured with 2 to 5 large nodules on each valve. Nodules prominent, in the radial midline and placed in alternate positions on each valve. Posterior ridge sharply defined, and posterior slope with numerous corrugations and small tubercles. Periostracum yellowish, yellowish brown, greenish brown or brown, and (in many pale specimens) with a broad greenish ray in the midline of each valve or with numerous very narrow rays (which may be broken into spots) that are generally distributed, or both. Nacre white, iridescent posteriorly, and in some specimens tinged with pink or blue. Beak sculpture fine, restricted to the umbonal apices, and composed of a few short, curved, oblique bars. Hinge teeth thick and strong: pseudocardinal teeth stumpy, elevated, deeply serrated, 1 or 2 in the right valve and 2 in the left; lateral teeth of medium length, well developed, 1 in the right valve and 2 in the left.

Unique in its combination of small size and few but heavy nodules that are restricted to a central radial row on each valve.

DISTRIBUTION

Lake Erie and its tributaries in Canada and the United States; Lake Michigan and its tributaries; throughout the Ohio-Mississippi drainage; and the Coosa-Alabama River system in the Gulf of Mexico drainage.

ECOLOGY

Occurs principally in large rivers and lakes. Recorded from substrates of gravel, sand, and mud. The breeding season is variously reported as short-term and long-term, and gravid specimens have been found in late spring and in summer. Glochidia are semicircular, without hooks, and measure about 0.22 mm in length and height. The host fish is unknown.

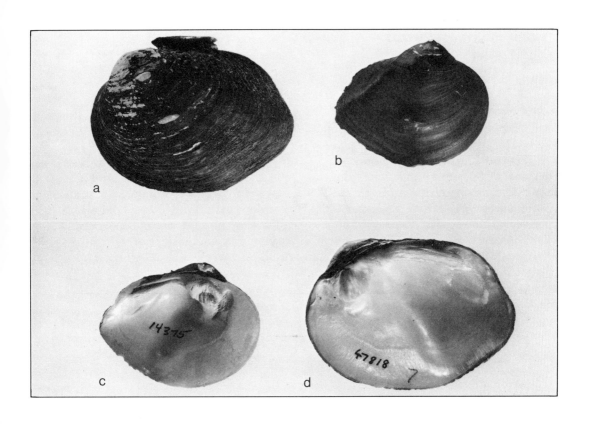

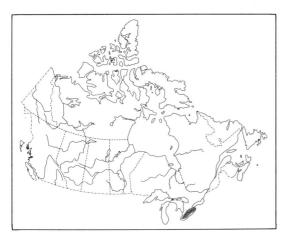

123
Obliquaria reflexa
a,d: South Bay, Pelee Is., L. Erie, Ont.; 57.2 mm.
b,c: St. Croix R., Hudson, Wisc.

The specimen illustrated in colour on page 239 is from the same locality as *a* and *d* (× 2/3).

124
Truncilla donaciformis
(Lea, 1828)
Fawn's-Foot

DESCRIPTION
Shell up to about 38 mm long, 25 mm high, 19 mm wide, and with mid-anterior shell wall about 2.5 mm thick; ovate, roundly pointed at mid-posterior, and rather thin but strong. Surface smooth except for elevated and rounded posterior ridge, fine concentric threads, and growth rests. Periostracum yellowish brown to brown, and (in many specimens) with numerous medium-sized rays that may be broken and joined to each other to form striking, irregular, zigzag, concentric bands. Nacre white, iridescent, and in some specimens with pale bluish or yellowish suffusions. Beak sculpture consisting of up to about 8 fine concentric bars that are straight or somewhat double-looped. Hinge teeth compressed, narrow, and well developed: pseudocardinal teeth sharp and well elevated, 1 in the right valve and 2 in the left (the posterior tooth is directly beneath the umbone); lateral teeth narrow, moderately long, 1 in the right valve and 2 in the left. The valves do not close tightly but gape posteriorly.

Resembles *T. truncata*, but differs in its smaller size, centrally pointed posterior margin, and rounded posterior ridge. It also lacks the diagonal posterior truncation of *T. truncata*.

DISTRIBUTION
In Canada known only from Lake Erie and the Grand River of southern Ontario. In the United States occurs in the Lake Erie and Lake Michigan watersheds, throughout much of the Ohio-Mississippi system, and in portions of the Gulf of Mexico drainage. Its limits in the latter area are uncertain because of its resemblance to another species, *T. macrodon* (Lea), which occurs there.

ECOLOGY
Occurs principally in rivers throughout its range. Found on sandy or muddy bottoms. Its breeding season is not fully determined, but gravid specimens have been reported from late spring to late summer. Glochidia are very small, semicircular, without hooks, and measure about 0.06 mm in length and height. The usual fish host is the freshwater drum, although the sauger is also utilized.

124
Truncilla donaciformis
a,d: Muskingum R. near Lowell, Washington Co., Ohio; 49.2 mm.
b,c: South Bay, Pelee Is., L. Erie, Ont.

The specimen illustrated in colour on page 239 is from the same locality as *a* and *d* (× 2/3).

125
Truncilla truncata
Rafinesque, 1820
Deer-Toe

DESCRIPTION
Shell up to about 70 mm long, 50 mm high, 35 mm wide, and with mid-anterior shell wall about 4 mm thick; roughly triangular or trapezoid-ovate, obliquely truncate posteriorly, but with posterior extremity bluntly pointed and, near the lower margin, moderately heavy and strong. Surface smooth except for the sharply rounded to keel-like posterior ridge, irregular concentric wrinkles, and well-marked growth rests. Periostracum yellowish brown, greenish or brown, and in most specimens with medium-sized greenish rays covering the shell. The rays are often broken into a series of spots or joined to each other to form a beautiful zigzag pattern. Nacre white, iridescent, and in some specimens with bluish or (more rarely) yellowish tints. Beak sculpture fine and consisting of 6 to 8 undulating bars that are in an open double-looped configuration and are confined to the umbonal apices. Hinge teeth well developed and compressed: pseudocardinal teeth sharp, serrated, elevated, 1 or 2 in the right valve and 2 in the left, of which the posterior tooth is located beneath the umbone; lateral teeth sharp and of medium length, 1 in the right valve and 2 in the left.

Well characterized by its more or less triangular shape, sharp posterior ridge, truncated and flattened posterior slope, diagonally truncated posterior margin, and unusual colour pattern. Compare with *T. donaciformis*.

DISTRIBUTION
Southern Ontario in the Lake Erie and Lake St. Clair drainages, central Great Lakes drainages in the United States, Ohio-Mississippi drainage throughout, and parts of the western Gulf of Mexico drainage.

ECOLOGY
Principally a river species like *T. donaciformis*. Found on gravel, sand, and mud. A long-term breeder; gravid specimens have been collected from May to August. Glochidia are very small, semicircular, and measure about 0.07 mm long and 0.08 mm high. The fish hosts are the freshwater drum and the sauger.

125
Truncilla truncata
a,d: Black Creek near Wilkesport, Ont.; 44.5 mm.
b,c: Grand R. near Pottawattomee Bayou, Mich.

The specimen illustrated in colour on page 240 is from the Grand R. near Dunnville, Ont. (× 2/3).

126
Proptera alata (Say, 1817)
Pink Heel-Splitter

DESCRIPTION
Shell up to about 150 mm long, 100 mm high, 55 mm wide, and with shell wall up to 12 mm thick at mid-anterior. Most adult specimens are close to 125 mm long, with shell wall about 6 mm thick. Shell ovate-trapezoid, with a prominent and triangular dorsal wing and with the posterior margin rounded or diagonally flattened, rather compressed in many specimens, and of moderate thickness and strength, especially anteriorly. Dorsal wing best developed in subadults. Shell sculptured with concentric wrinkles and growth rests, and with a low single or double posterior ridge placed high on the shell. Periostracum greenish brown to brown in young specimens, brown to blackish in adults, and with or without poorly defined rays. Nacre purple (usual) to pink and iridescent. Beak sculpture composed of 4 or 5 short single- or slightly double-looped low bars close to the umbonal apices. Hinge teeth well developed: pseudocardinal teeth conical, serrated, of medium size, 1 or 2 in the right valve and 2 in the left; lateral teeth elevated, elongate, curved, 1 in the right valve and 2 in the left. Female specimens are slightly more swollen than males but the difference is slight.

Easily distinguished by its large size, prominent posterior wing, dark periostracum, strong and complete hinge teeth, and purple nacre. The other large alate species, *Lasmigona complanata* and *Leptodea fragilis*, differ in that the former has white nacre and no (or vestigial) lateral teeth, and the latter has a yellowish periostracum, weak hinge teeth, pinkish rather than purple nacre, and a fragile shell.

DISTRIBUTION
In Canada occurs in the Red River and Winnipeg River systems of the Canadian Interior Basin and in the St. Lawrence River system from Lake St. Clair and its tributaries to Lake Champlain and the vicinity of Montreal. In the United States, in addition to the Great Lakes–St. Lawrence system (except Lake Superior), it occurs throughout the Ohio-Mississippi system and the Gulf of Mexico drainage from Alabama to Texas.

ECOLOGY
Found in large rivers, in lakes, and in canals, usually in mud. A long-term breeder (August to July). The glochidia are unusual; they are axe-head shaped, with two spines on each valve, and measure about 0.22 mm in length and 0.40 mm in height. The only known host fish is the freshwater drum.

126
Proptera alata
a,d: Grand R., Dunnville, Ont.; 139.7 mm.
b,c: Rondeau Harbour; L. Erie, Ont.

The specimen illustrated in colour on page 240 is from Rondeau Harbour, L. Erie, near Shrewsbury, Ont. (× 2/3).

127
Carunculina parva
(Barnes, 1823)
Lilliput Mussel

DESCRIPTION
Shell up to about 30 mm long, 18 mm high, 15 mm wide, and with shell wall about 1.5 mm thick at mid-anterior; elliptical and moderately inflated in males, ovate and more inflated in females, evenly rounded posteriorly, and thin but strong. Shell unsculptured except for the beak sculpture and the low posterior ridge placed high on the shell. Periostracum folded in lines of growth, cloth-like, brown to brownish black, and without rays. Nacre iridescent and silvery white or bluish. Beak sculpture heavy and consisting of about 6 oblique and irregular curved bars. Hinge teeth compressed but fully developed: pseudocardinal teeth conical, erect, serrated, 1 (sometimes split) in the right valve and 2 in the left; lateral teeth nearly straight, 1 in the right valve and 2 in the left.

The small size, elliptical to ovate shape, rounded posterior, cloth-like and rayless periostracum, oblique beak sculpture, and compressed but complete and normal hinge teeth distinguish this species from all others. Compare with *Alasmidonta heterodon*, *Simpsoniconcha ambigua*, and *Villosa fabalis*.

DISTRIBUTION
Great Lakes-St. Lawrence system in tributaries of Lake Erie (Canada and the United States); Ohio-Mississippi system throughout; and the Apalachicolan region of the Gulf of Mexico drainage.

ECOLOGY
Usually found on muddy bottoms in sluggish streams and canals. Probably a long-term breeder, although gravid specimens have been found only in late spring and in summer. Some populations are reported to contain only specimens with female gill structure, but others contain specimens of both male and female gill types. Glochidia are roughly elliptical, without hooks, and measure about 0.18 mm in length and 0.20 mm in height. The host fish are green sunfish, warmouth, orange spotted sunfish, bluegill, and white crappie.

Females of all species of *Carunculina* have a caruncle (a thickened, knob-like protuberance) on the posterior ventral edge of the mantle.

127
Carunculina parva
a,d: Ouachita R., Arkadelphia, Ark.
b,c: Salt Fork, Vermillion R., Homer Park, Ill.; 31.6 mm.

The specimen illustrated in colour on page 240 is from McGregor Creek, Chatham, Ont. (× 2/3).

128
Obovaria olivaria
(Rafinesque, 1820)
Olive Hickory-Nut

DESCRIPTION

Shell up to about 55 mm long, 40 mm high, 25 mm wide, and with mid-anterior shell wall 5 mm thick; shell ovate, thick, and strong. Some southern specimens are more than 75 mm long, with shell wall nearly 8 mm thick. Surface smooth and unsculptured except for fine and crowded concentric lines near the margin, and growth rests. Periostracum olive-green or modified with yellow or brown, and covered with mostly narrow green rays. When visible the rays are most apparent in the central area of each valve. Nacre silvery white or faintly bluish. Beaks near or at the anterior end, swollen and elevated, incurved, directed forward, and badly worn in most specimens. Beak sculpture rudimentary and consisting of 4 or 5 fine, centrally sinuate bars. Hinge teeth rather heavy and strong: pseudocardinal teeth thick and stubby or slightly lengthened and parallel with lateral teeth, 1 in the right valve and 2 in the left; lateral teeth moderately long, heavy, slightly curved, 1 or 2 in the right valve and 2 in the left.

Easily recognized by its ovate shape, evenly rounded outline without angles, swollen and anterior beaks, olive-green colour, and relatively thick hinge teeth. Compare with *O. subrotunda*.

DISTRIBUTION

Available Canadian specimens are all from the St. Lawrence River and two of its tributaries, the Ottawa and St. Francis rivers. In the United States the species is reported from a few widely scattered localities in the Lake Ontario, Lake Erie and Lake Huron drainages and from many localities throughout the Ohio-Mississippi system.

ECOLOGY

Occurs chiefly in large rivers. Found on sandy bottoms in moderate to rather deep water. A long-term breeder; its gravid period extends from August to the following June. Glochidia are oval, without hooks, and measure about 0.19 mm in length and 0.22 mm in height. The freshwater drum and the shovelnose sturgeon are host fish.

128
Obovaria olivaria
a,d: St. Lawrence R. near Montmorency, Que.; 54.0 mm.
b,c: Ottawa R. near Ottawa, Ont.

The specimen illustrated in colour on page 240 is from the Ottawa R. at MacLarens Landing, Ont. (× 2/3).

129
Obovaria subrotunda
(Rafinesque, 1820)
Round Hickory-Nut

DESCRIPTION
Shell up to about 65 mm long, 50 mm high, 40 mm wide, and with shell wall 6.3 mm thick at mid-anterior. Most specimens are smaller and measure about 40 mm long and 4 mm in shell-wall thickness. Shell circular to triangular-circular, thick, and strong. Surface smooth except for prominent growth rests. Periostracum yellowish brown, greenish brown or brown, and without rays. Nacre silvery white or with a faint flush of blue or pink. Beaks moderately inflated, incurved, and located centrally or a little anteriorly. Beak sculpture fine and consisting of about 6 nearly straight, short bars. Hinge teeth rather heavy and strong: pseudocardinal teeth erect and conical, 3 in the right valve (the central tooth large, the others small) and 2 in the left; lateral teeth rather short, thick, and nearly straight, 1 in the right valve and 2 in the left.

The circular shape, unsculptured and unrayed surface, medium-to-small size, centrally located beaks, heavy shell, and white nacre readily distinguish this species. It differs from *O. olivaria* in shape, relative position of the beaks, and colour pattern.

DISTRIBUTION
Lake Erie and Lake St. Clair and their drainages in Canada and the United States, and the Ohio-Mississippi River system throughout.

ECOLOGY
Typically a river species living on sandy bottoms, but also occurs in Lake Erie. A long-term breeder, its gravid period extends from about September to June. Glochidia are ovate, with a nearly straight hinge line, without hooks, and measure about 0.20 mm long and 0.23 mm high. The host fish is unknown.

129
Obovaria subrotunda
a,d: Sydenham R. near Strathroy, Ont.; 58.7 mm.
b,c: Sydenham R. near Alvinston, Ont.

The specimen illustrated in colour on page 241 is from South Bay, Pelee Is., L. Erie, Ont. (× 2/3).

130
Leptodea fragilis
(Rafinesque, 1820)
Fragile Paper-Shell

DESCRIPTION

Shell up to about 150 mm long, 90 mm high, 45 mm wide, and with mid-anterior shell wall about 3 mm thick; ovate-trapezoid, thin, fragile, compressed, and with a prominent triangular dorsal wing. Dorsal wing high in juveniles, low in old adults. Shell smooth except for low concentric wrinkles, growth rests, and low posterior ridge placed high on the shell. Periostracum pale yellow to yellowish brown, with dark growth rests and, in some specimens, poorly defined narrow green rays. Nacre silvery white and iridescent or pinkish on some specimens, especially near the hinge. Beak sculpture fine, completely visible only on very young specimens, and composed of 4 or 5 double-looped bars with elevated lower apices that form 2 short radial rows of nodules. Hinge teeth narrow: pseudocardinal teeth thin, weak, 1 or a rudimentary tooth in the right valve, and 1 or 2 or a rudimentary tooth in the left; lateral teeth long, thin, elevated, 1 in the right valve and 2 less elevated in the left.

Distinguished by its dorsal wing, thin and fragile shell, large size, yellowish periostracum, whitish nacre, and narrow hinge teeth. Compare with *Proptera alata* and *Lasmigona complanata*.

DISTRIBUTION

Entire Great Lakes-St. Lawrence system in Canada and the United States from Lake Michigan and its tributaries to tidewater in the St. Lawrence River (near Quebec City); Ohio-Mississippi system; and western Gulf of Mexico drainage.

ECOLOGY

Occurs in rivers of various widths, lakes, and canals. Found on bottoms of mud, sand, or gravel. A long-term breeder, with gravid period extending from August to July. Glochidia are ovate, very small, and measure about 0.08 mm in length and 0.09 mm in height. The host fish is the freshwater drum.

130
Leptodea fragilis
a,d: Grand R., Byng, Ont.; 133.4 mm.
b,c: L. Erie at Point Pelee, Ont.

The specimen illustrated in colour on page 241 is from the Grand R. near Cayuga, Ont. (× 2/3).

131
Actinonaias carinata
(Barnes, 1823)
Mucket

DESCRIPTION
Shell up to about 150 mm long, 80 mm high, 65 mm wide, and with mid-anterior shell wall 10 mm thick below the pallial line; ovate or ovate-elliptical, thick, strong, and a little inflated but somewhat flattened centrally. Surface unsculptured except roughened by concentric wrinkles and lines of growth. Periostracum yellowish, greenish, or brownish and with broad rays in many specimens, especially juveniles. Nacre white, very rarely pinkish, ordinarily covered with very tiny elevations (pimples), and with a narrow greenish or brownish band bordering the margin. Beaks compressed, not inflated, and projecting only slightly above hinge plate; beak cavity narrowly excavated. Beak sculpture poorly developed and consisting of a few narrow concentric bars that are single- or slightly double-looped. Hinge teeth well developed and heavy in adults: pseudocardinal teeth sub-triangular, elevated, 1 or 2 in the right valve and 2 in the left; lateral teeth long and continuing as ridges nearly to the umbones, 1 in the right valve and 2 in the left. Sexual dimorphism in shells not apparent.

Easily differentiated from all other species except males of *Lampsilis ventricosa*. Distinguished from *L. ventricosa* by its low beaks that are not elevated much above the ligament, narrowly excavated beak cavities, centrally thickened and flattened shell, nacre surface that appears nearly flat or slightly convex below the beak cavity, and apparent lack of sexual dimorphism.

DISTRIBUTION
Occurs in tributaries of all of the Great Lakes except Lake Superior, but ordinarily rare. Abundant throughout the Ohio-Mississippi River system.

ECOLOGY
Rare in Canada. A river species usually found in gravel or sand. A long-term breeder with gravid period extending from August to May. Glochidia are more or less elliptical, without hooks, and measure about 0.22 mm in length and 0.25 mm in height. Several fish species are hosts for the glochidia.

131
Actinonaias carinata
a,d: Sydenham R. near Shetland, Ont.; 133.4 mm.
b,c: Thames R., Chatham, Ont.

The specimen illustrated in colour on page 241 is from the same locality as *a* and *d* ($\times$ 2/3).

132
Ligumia nasuta (Say, 1817)
Pointed Sand-Shell

DESCRIPTION

Shell up to about 100 mm long, 45 mm high, 25 mm wide, and with mid-anterior shell wall 3.5 mm thick; long-elliptical, with posterior centrally pointed, rather compressed, and thin but strong. Surface with concentric wrinkles and clearly visible lines of growth. Posterior ridge well marked, angular near the beaks, rounded distally, and terminating at the posterior point. Periostracum olive-green, brown or blackish, and with narrow rays in some specimens. Nacre purple, salmon, or silvery white with or without interior yellowish suffusions. Beaks pointed, compressed, and projecting slightly above hinge line. Beak sculpture composed of about 8 rather fine, double-looped concentric bars. Hinge teeth delicate: pseudocardinal teeth small, erect, compressed, 1 or 2 in each valve; lateral teeth narrow, long, 1 in the right valve and 2 in the left. Females are more expanded post-basally than males.

Well characterized by its moderately sized narrow shell, centrally pointed posterior, absence of prominent rays, and (when present) purple nacre. Purple nacre is dominant in populations from the Atlantic drainage but rare in those from the Great Lakes drainage.

DISTRIBUTION

In Canada occurs only in Lake Ontario, Lake Erie, Lake St. Clair and their drainage basins. In the United States it occurs in those same lakes and drainage basins, from Michigan to New York, but also farther east to the Atlantic drainage in Massachusetts, and south in the Atlantic drainage to the James River in Virginia.

ECOLOGY

Occurs in protected areas of lakes, in slack-water areas of rivers, and in canals. Found in mud or sand. Breeds from August until the following June. Glochidia are subovate, with an undulate hinge line, and measure about 0.25 mm long and 0.29 mm high. The host fish has not been determined.

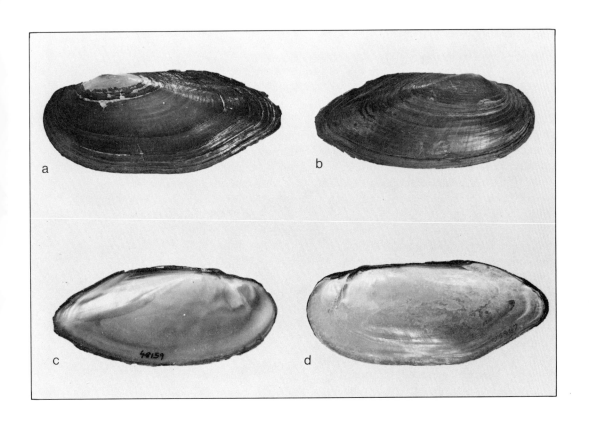

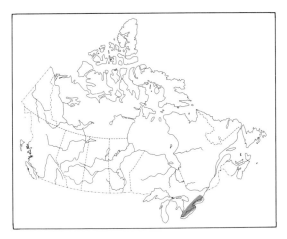

132
Ligumia nasuta
a,d: Bay of Quinte, L. Ontario, Ont.; 87.3 mm (female).
b,c: Big Creek near Port Rowan, Ont. (male).

The female specimen illustrated in colour on page 242 is from Winnekonnet Pond near Norton, Bristol Co., Mass. (× 2/3).

133
Ligumia recta (Lamarck, 1819)
Black Sand-Shell

DESCRIPTION
Shell up to about 175 mm long, 75 mm high, 45 mm wide (males) and with mid-anterior shell wall about 8 mm thick. Females are proportionately broader, especially posteriorly. Shell long-elliptical, roundly pointed posteriorly in males, evenly rounded posteriorly in females, thick, and strong. Surface smooth except for concentric wrinkles and growth rests. Periostracum dark green, dark brown or nearly black, and with darker rays that are obscure in adults. Nacre silvery white throughout, or partially to wholly purple or pink. Beak sculpture faint, obsolete, and consisting of 3 to 5 indistinct double-looped bars that disappear on the posterior slope. Hinge teeth well developed and strong: pseudocardinal teeth conical, elevated, serrated, 1 or 2 in the right valve and 2 in the left; lateral teeth long, also elevated, 1 in the right valve and 2 in the left.

Easily recognized by its large size, lancehead shape, well-developed hinge teeth, and almost black colour. Compare with *L. nasuta*.

DISTRIBUTION
Hudson Bay drainage in the Red River and Winnipeg River systems and Lake Winnipeg; Great Lakes-St. Lawrence system in all the Great Lakes except Lake Superior and many of their tributaries, east to the Ottawa River, Lake Champlain, and the St. Lawrence River to the vicinity of Montreal; Ohio-Mississippi system throughout; and Alabama River system.

ECOLOGY
Occurs principally in large rivers but is also found in canals and large lakes. Usual substrates are sand or gravel; occasionally found in mud. A long-term breeder; its gravid period overlaps in different individuals and extends from June, July and August through the winter to the following summer. The glochidia are purse-shaped, ovate, and measure about 0.23 mm long and 0.27 mm high. Host fish are the American eel, bluegill, largemouth bass, and white crappie.

133
Ligumia recta
a,d: Ottawa R., Pointe-aux-Anglais, Que.; 120.7 mm (male).
b,c: Same locality (female).

The female specimen illustrated in colour on page 242 is from the Sydenham R. near Shetland, Ont. (× 2/3).

134
Lampsilis cariosa (Say, 1817)
Yellow Lamp-Mussel

DESCRIPTION

The largest Canadian specimen is 100 mm long, 70 mm high, 45 mm wide, and with mid-anterior shell wall 4 mm thick. Specimens from some localities farther south may be up to 140 mm long and have relatively thick shells. Shell evenly ovate in males and ovate but higher posteriorly than anteriorly in females, moderately thick and strong, and somewhat inflated. Surface smooth except for concentric growth rests. Periostracum ordinarily bright orange-yellow, more rarely yellowish brown or reddish brown, and without rays or with narrow rays only on, or close to, the posterior slope. Nacre predominantly white and (in some specimens) suffused with pink or orange in the beak cavities or posteriorly. Beaks moderately inflated and projecting above hinge line, and beak cavities well excavated. Beak sculpture in most specimens obscured by corrosion, but composed of 5 or 6 moderately coarse, curved, single- or slightly double-looped concentric bars. Hinge teeth complete and well developed: pseudocardinal teeth elevated, more or less conical, somewhat compressed (but not lamellate) and directed forward, 2 in each valve; interdentum thick and rather heavy; lateral teeth elevated, of medium length, straight or a little curved, 1 in the right valve and 2 in the left.

Similar to *L. ventricosa* but normally smaller, with predominantly yellowish rather than predominantly brownish periostracum; when rays are present they are narrow and restricted to the vicinity of the posterior slope. Compare also with *L. ochracea*.

DISTRIBUTION

Atlantic drainage from Sydney River on Cape Breton Island to the Ogeechee River system, Georgia.

ECOLOGY

Predominantly a river species. It occurs chiefly in rather swift currents on shoals or in riffles and principally on sand bottoms. Occasionally it is found in ponds. A long-term breeder but the limits of its breeding season are not known. Hermaphrodite specimens have been reported. Glochidia are roughly elliptical, with a straight hinge line, without hooks, and measure about 0.22 mm in length and 0.28 mm in height. The host fish is unknown.

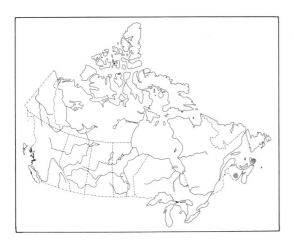

134
Lampsilis cariosa
a,d: Sydney R. near Sydney, N.S.; 88.9 mm (male).
b,c: Merrimac R., Haverhill, Essex Co., Mass. (female).

The female specimen illustrated in colour on page 242 is from the same locality as *b* and *c* (× 2/3).

135
Lampsilis fasciola
Rafinesque, 1820
Wavy-rayed Lamp-Mussel

DESCRIPTION
Shell up to about 95 mm long, 65 mm high, 45 mm wide, and with mid-anterior shell wall about 7.5 mm thick; quadrate-ovate (males) or ovate (females), heavy and strong, moderately inflated, and heavily rayed. Surface smooth except for concentric wrinkles and growth rests. Posterior ridge indistinct. Periostracum yellowish, greenish yellow or yellowish brown, and covered with crowded, narrow and wide, interrupted, wavy rays. Many of the wide rays are composed of closely aligned, very narrow rays. Nacre white or bluish white. Beaks elevated, and beak cavities moderately excavated. Beak sculpture rather fine and composed of about 6 concentric broadly curved bars that are sinuous or broken in the centre. Hinge teeth well developed and moderately heavy: pseudocardinal teeth stumpy or subconical, elevated, serrated, 2 in the right valve (the anterior tooth small) and 2 in the left; lateral teeth rather short, strong, slightly curved, 1 in the right valve and 2 in the left.

Somewhat similar to *L. ventricosa* but is smaller, relatively thicker, and more regularly ovate. The chief difference is in the character of the rays, which are narrow and individual, or narrow and coalesced into wide rays, but are always wavy and with multiple interruptions. In *L. ventricosa*, rays are not wavy and have only a few interruptions.

DISTRIBUTION
Great Lakes drainage in the tributaries of Lake Michigan, Lake Huron, Lake St. Clair and Lake Erie, and Ohio-Mississippi drainage south to the Tennessee River system.

ECOLOGY
Uncommon in Canada. Occurs in rivers, usually in regions of riffles, and on gravel or sand bottoms. Its long-term breeding season extends from early August to the following July and August. Glochidia are purse-shaped, without spines, and measure about 0.24 mm in length and 0.29 mm in height. The host fish has not been determined.

135
Lampsilis fasciola
a,d: Grand R., West Montrose, Ont.; 68.3 mm (female).
b,c: Raisin R., Jackson-Washtenaw Co. line, Mich. (male).

The specimen illustrated in colour on page 243 is from the Huron R. near Manchester, Washtenaw Co., Mich. (× 2/3).

136
Lampsilis ochracea (Say, 1817)
Delicate Lamp-Mussel

DESCRIPTION

Shell up to about 80 mm long, 55 mm high, 40 mm wide, and with anterior ventral shell wall about 4 mm thick near the pallial line. Most specimens are somewhat smaller. Shell ovate in both sexes (males relatively longer with bluntly pointed posterior, females with rounded posterior), centrally inflated, rather thin, and of only moderate strength. Surface with low concentric wrinkles and prominent growth rests. Periostracum brownish or modified with greenish, reddish or yellowish shades, and without rays or with narrow, greenish, rather obscure rays all over the shell. Nacre predominantly white but also bluish white, or tinted with pink or salmon. Beaks moderately inflated and projecting above hinge line; beak cavities excavated. Beak sculpture composed of about 6 medium-sized, curved, concentric, single-looped bars. Hinge teeth complete but markedly compressed and narrow: pseudocardinals elevated, compressed, lamellate and directed forward, 2 in each valve or rarely 3 in the left valve; interdentum thin and compressed; lateral teeth thin, a little curved, 1 in the right valve and 2 in the left.

Resembles some specimens of *L. cariosa* but is typically smaller and more delicate, with rays either absent or covering the shell, with lamellate pseudocardinals, and with a narrow and compressed interdentum. *L. cariosa* has rays only near the posterior slope, compressed but not lamellate pseudocardinals, and a thick and rather heavy interdentum. Also resembles some specimens of *L. radiata radiata*, but in that subspecies sexual dimorphism is not apparent and the pseudocardinal teeth are not lamellate.

DISTRIBUTION

Atlantic coastal plain from Cape Breton, Nova Scotia, to the Savannah River, Georgia.

ECOLOGY

Occurs principally in quiet water, that is ponds, canals, and slow-moving parts of rivers, and is therefore unlike *L. cariosa*. Found on mud or sand bottoms. Occurs only near the seacoast. A long-term breeder, but limits of its gravid period and details of its glochidia are unknown. Its fish host is also unknown, but its habitat suggests an anadromous fish, possibly the alewife.

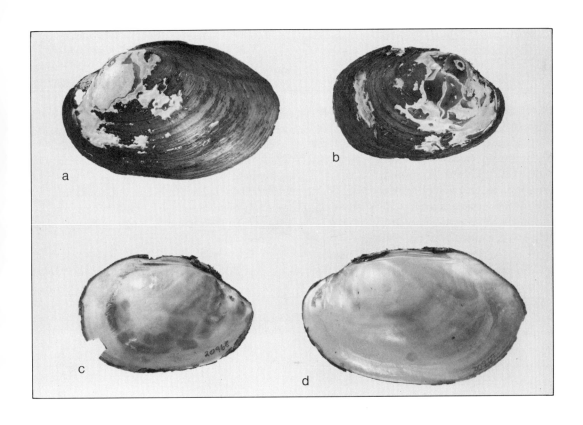

136
Lampsilis ochracea
a,d: West branch of Aulac R. near Sackville, N.B.; 71.4 mm (male).
b,c: Same locality (female).

The specimen illustrated in colour on page 243 is from the Sydney R. near Sydney, N.S. (× 1).

137
Lampsilis radiata radiata
(Gmelin, 1792)
Eastern Lamp-Mussel

DESCRIPTION
Shell up to about 105 mm long, 65 mm high, 40 mm wide, and with shell wall about 6 mm thick at mid-anterior; ovate-elliptical, not inflated, without prominent sexual dimorphism, and of moderate thickness and strength; covered, especially in young specimens, with numerous, mostly broad rays. Shell with rounded but poorly defined posterior ridge, well-marked growth rests, and fine concentric threads. Periostracum yellowish brown, brown or blackish, and rayed in most individuals, but rays obscure in old, blackened specimens. Nacre white to bluish white or pinkish. Beaks low and barely projecting above hinge line; beak cavities shallow. Beak sculpture not visible in most specimens, but in very young specimens about 6 concentric, open, double-looped bars can be seen. Hinge teeth medium-sized and well developed: pseudocardinal teeth erect, pyramidal, serrate, 2 or 3 in the right valve and 2 in the left; lateral teeth well developed, straight or slightly curved, 1 in the right valve and 2 in the left.

The ovate shape, medium size, wide and generally distributed rays in young specimens, complete hinge teeth, and generally white nacre will distinguish this species from all others in the Atlantic drainage. In the St. Lawrence system it intergrades with *L. r. siliquoidea* (Barnes), with which it should be compared.

DISTRIBUTION
Lower St. Lawrence system and south in the Atlantic drainage to the Pee Dee River system in South Carolina.

ECOLOGY
A common species. Occurs in rivers and lakes of all sizes, usually on gravel and sand bottoms but occasionally on mud. The long-term breeding season begins about August and ends the following August. Glochidia are oval, without hooks, and measure about 0.23 mm in length and 0.28 mm in height. The host fish is unknown.

137
Lampsilis radiata radiata
a,d: L. Champlain, Sand Bar State Park near Burlington, Vt.; 88.9 mm (male).
b,c: Same locality (female).

The specimen illustrated in colour on page 243 is from Darlings L. near Lakeside, King's Co., N.B. (× 2/3).

138
Lampsilis radiata siliquoidea
(Barnes, 1823)
Fat Mucket

DESCRIPTION

Shell up to about 140 mm long, 70 mm high, 55 mm wide, and with anterior ventral shell wall up to 12 mm thick near pallial line. Most specimens are much smaller. Shell elliptical, expanded in females (but not in males) in the posterior part of the ventral margin, heavy, strong, and inflated. Surface roughened by concentric wrinkles and lines of growth. Periostracum yellowish, greenish or brownish, shiny, and in most specimens covered by extensive, sharply defined, narrow rays. Nacre white or bluish white and iridescent posteriorly. Beaks low and projecting only a little above hinge line; beak cavities shallow. Beak sculpture rather coarse and consisting of numerous concentric bars that have a shallow central sinuation or are centrally broken. Hinge teeth well developed and fairly strong: pseudocardinal teeth medium in size, erect, serrated, compressed, directed forward, 2 in each valve; lateral teeth narrow, prominent, straight or slightly curved, 1 in the right valve and 2 in the left.

Closely related to only *L. r. radiata*, but may be distinguished as follows: exhibits prominent sexual dimorphism, has shiny yellowish-brown to brown periostracum, and bears numerous narrow rays; *L. r. radiata* exhibits vague or no sexual dimorphism, has cloth-like dark-brown to blackish periostracum, and bears numerous wide rays. Attains a larger size than *L. r. radiata* and does not have the pink nacre frequently found in some *L. r. radiata* populations. In the Lake Ontario and St. Lawrence River watersheds intermediate populations occur, however, and these are most conveniently referred to as *Lampsilis radiata* (*sensu lato*).

DISTRIBUTION

Canadian Interior Basin from Quebec to Alberta; Mackenzie River system north to the vicinity of Great Slave Lake; Great Lakes drainage from the Lake Superior to Lake Erie watersheds (see mention of intergrades above); and upper Ohio-Mississippi drainage from New York to Minnesota and Arkansas.

ECOLOGY

Abundant. Occurs in rivers and lakes of all sizes. Found on all types of bottoms (clay, mud, sand, or gravel). Often lives in river banks in water as shallow as 5 to 8 cm. A long-term breeder, with gravid period extending from the first part of August to the middle of the following July. Glochidia are purse-shaped, without hooks, and measure from 0.24 to 0.26 mm in length and 0.26 to 0.30 mm in height. The host fish are black crappie, bluegill, largemouth bass, rock bass, sauger, smallmouth bass, white bass, white crappie, yellow perch, and yellow pike-perch.

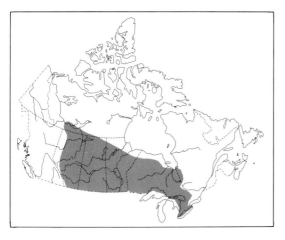

138
Lampsilis radiata siliquoidea
a,d: Assiniboine R., Winnipeg, Man.; 69.9 mm (male).
b,c: Mouth of Hay R., Hay River, N.W.T. (female).

The specimen illustrated in colour on page 243 is from Flat Creek southeast of Blenheim, Ont. (× 2/3).

139
Lampsilis ventricosa
(Barnes, 1823)
Pocket-Book

DESCRIPTION
Old specimens may be up to about 155 mm long, 100 mm high, 75 mm wide, and with anterior ventral shell wall about 13 mm thick below pallial line. Most specimens are considerably smaller. Shell ovate to elliptical, highest centrally in males and highest posteriorly in females, massive and strong in old specimens from hard water, inflated, and with exterior evenly convex centrally. Surface unsculptured except roughened by concentric wrinkles and lines of growth. Periostracum yellowish to olive-brown and typically with narrow-to-wide dark-green rays covering the entire shell. Rays may be nearly or entirely lacking, especially in old specimens. Nacre white or bluish white. Beaks inflated and projecting well above hinge plate; beak cavity deeply excavated. Beak sculpture coarse and composed of 5 or 6 concentric, weakly double-looped bars, the last 2 or 3 prominent. Hinge teeth well developed, thick, and moderately heavy: pseudocardinal teeth prominent, elevated, conical, compressed, and directed forward, 2 in each valve; lateral teeth strong, of medium length, not continuing close to the umbones, 1 in the right valve and 2 in the left. Sexual dimorphism prominent.

Clearly distinct from all other species except *Actinonaias carinata*. Compare with that species.

DISTRIBUTION
Winnipeg, Red, and Nelson River systems of central Canada; Great Lakes-St. Lawrence system throughout (except most of Lake Superior); Ohio-Mississippi River system throughout; and Potomac River system (introduced).

ECOLOGY
Usually found in rivers but occasionally in lakes. Substrates inhabited are of all types. A long-term breeder, with gravid period extending from the end of July to near the beginning of the following July. Glochidia are ovate, with nearly straight hinge line, and measure about 0.25 mm long and 0.30 mm high. The posterior edge of the female mantle protrudes from the shell and resembles a fish (see frontispiece). Known hosts for the glochidia are bluegill, largemouth bass, smallmouth bass, white crappie, yellow perch, and yellow pike-perch.

Recent work at the Ohio State University Museum of Zoology has shown that *L. ventricosa* and *L. ovata* (Say, 1817) are biologically distinct species. The Canadian mussel formerly called *Lampsilis ovata ventricosa* should therefore be called *Lampsilis ventricosa*.

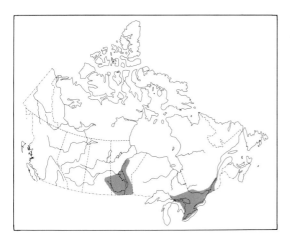

139
Lampsilis ventricosa
a,d: Ottawa R., Oka, Que.; 87.3 mm (male).
b,c: Same locality (female).

The female specimen illustrated in colour on page 244 is from the Sydenham R. northeast of Shetland, Ont. (× 2/3).

140
Villosa fabalis (Lea, 1831)
Bean Villosa

DESCRIPTION
Shell up to 38 mm long, 19 mm high, 13 mm wide, and with mid-anterior shell wall 2.5 mm thick; sub-elliptical, very small, and solid. Females are generally more inflated and more broadly rounded posteriorly than males. Surface with low concentric lines and wrinkles and dark growth rests. Periostracum normally light or dark green and covered with darker green rays. Rays wide or narrow, wavy, and clearly apparent except on old, blackened specimens. Nacre silvery white and iridescent. Beaks narrow, slightly elevated above hinge line, and not excavated. Beak sculpture fine and composed of about 5 crowded double-looped ridges. Hinge teeth relatively heavy: pseudocardinals erect, pyramidal, serrated or ragged, 1 in the right valve and 2 in the left; interdentum thick; laterals short, elevated, straight or a little curved, with diagonal serrations, 1 in the right valve and 2 in the left.

Easily distinguished by its very small size, elliptical shape, crowded rays, and heavy hinge teeth. The other diminutive species, *Carunculina parva*, is more inflated, more rounded posteriorly, lacks rays, and has narrow hinge teeth. Compare also with *Alasmidonta heterodon*.

DISTRIBUTION
In Canada known only from the Sydenham River (Lake St. Clair drainage) in southwestern Ontario. In the United States it occurs in Lake Erie, in tributaries of Lake Michigan, and in the Ohio-Mississippi system from the Allegheny River system to the Tennessee River system.

ECOLOGY
A rare species. Occurs in riffle areas of small rivers among the roots of submersed vegetation. Limits of the breeding season are not known but gravid specimens have been found in May. Probably a long-term breeder. Glochidia are rounded but with a straight hinge line, and measure about 0.17 mm in length and 0.20 mm in height. The host fish is unknown.

140
Villosa fabalis
a,d: "Ohio"; 34.9 mm.
b,c: East Fork, Little Miami R. near Williamsburg, Ohio.

The specimen illustrated in colour on page 244 is identified only as being from "Ohio" (× 2/3).

141
Villosa iris (Lea, 1830)
Rainbow Shell

DESCRIPTION
Shell up to about 75 mm long, 45 mm high, 30 mm wide, and with anterior shell wall 4 mm thick near pallial line. Most specimens are smaller and relatively narrower. Shell long-elliptical, somewhat pointed posteriorly, and rather small. Surface smooth except for low concentric wrinkles, fine concentric lines, and well-marked growth rests. Periostracum yellowish, pale green, or brown (in old specimens), and with wide, or both narrow and wide, dark-green rays covering the whole surface (or absent posteriorly) and with concentric pale lines and bands that interrupt the rays in many specimens. In old, brown specimens the rays may be obscure. Nacre silvery white and iridescent. Beaks far forward, narrow and only slightly elevated above hinge line; beak cavities shallowly excavated. Beak sculpture distinct and consisting of 4 to 6 medium-sized bars, the first bar subconcentric and the following bars double-looped or irregular and nodulous. Hinge teeth medium-sized, well developed, and complete: pseudocardinals elevated, a little compressed, conical, serrated, 1 major tooth in the right valve and 2 in the left; laterals rather long and narrow, 1 in the right valve and 2 in the left.

May be recognized by its rather small size, narrow form, and prominent, wide interrupted rays. Compare with *V. fabalis*.

DISTRIBUTION
Great Lakes system in Lakes Michigan, Huron, St. Clair, Erie, Ontario and their tributaries; Ohio-Mississippi system south to, and including, the Tennessee River drainage area.

ECOLOGY
Occurs in lakes, rivers, and medium-sized creeks where moderate currents prevail. Found on sand or gravel bottoms. Hermaphroditic. The terminal dates of its long-term breeding season are not known. Glochidia are semi-elliptical, large, with a short hinge line, and measure about 0.23 mm in length and 0.29 mm in height. The host fish has not been discovered.

141
Villosa iris
a,b,c,d: Fairchild Creek near Cainsville, Ont.; *a* and *d* 61.9 mm.

The specimen illustrated in colour on page 244 is from L. Erie near Rondeau Park, Ont. (× 2/3).

142
Dysnomia torulosa rangiana
(Lea, 1839)
Northern Riffle Shell

DESCRIPTION
A mature male is 45 mm long, 35 mm high, 21 mm wide, and has a mid-anterior shell-wall thickness of 4 mm. The shell is ovate except biangulate posteriorly and with indented posterior ventral margin, of medium inflation, with a centrally located rounded radial ridge, a rounded posterior ridge, a shallow furrow between the ridges, and with maximum height near the middle. A mature female is nearly 50 mm long, 35 mm high, 24 mm wide, and also has a mid-anterior shell-wall thickness of 4 mm. The shell is quadrate-ovate, broadly and roundly swollen, thin shelled in the posterior basal region, without radial ridges, and with maximum height near the posterior third of the shell. In both sexes the periostracum is yellowish or greenish olive, and has generally distributed, crowded, uninterrupted, broad or narrow, dark-green rays and dark growth rests. In many females the rays are absent on the posterior expansion. Nacre white anteriorly, bluish white posteriorly, and iridescent. In females nacre is nearly lacking in a 6-mm-wide band at the edge of the thin-walled posterior expansion. Beaks elevated above hinge line and moderately excavated. Beak sculpture fine and consisting of a few gently curved, single-looped bars. Hinge teeth medium-sized and well developed: pseudocardinals conical, deeply serrated, 2 in the right valve (the anterior one small) and 2 in the left; laterals of medium length, elevated, straight, blade-like, 1 in the right valve and 2 in the left.

Unmistakable among Canadian unionids because of its extreme and unique sexual dimorphism and small size. Females are expanded ventrally, that is distended along the posterior ventral margin, whereas females of *D. triquetra*, the other Canadian species of *Dysnomia*, are expanded laterally, that is are swollen along the posterior ridge and are sharply truncated posteriorly.

DISTRIBUTION
In Canada is known to live only in the Sydenham River (Lake St. Clair drainage) in southwestern Ontario. In the nineteenth century it was also collected from the Canadian side of Lake Erie. In the United States is also known from Lake Erie and some of its tributaries and from the Ohio River system. Farther south it is replaced by other subspecies of *D. torulosa*. Like most other species and subspecies of *Dysnomia*, populations of this very rare unionid have been devastated by damming and pollution throughout its original range.

ECOLOGY
Rare. Lives principally in highly oxygenated riffle areas of rivers. Found on rocky and sandy bottoms. A long-term breeder; its gravid period probably extends from late summer to the following spring. Glochidia are semicircular, with straight hinge line and without hooks. The host fish is not known.

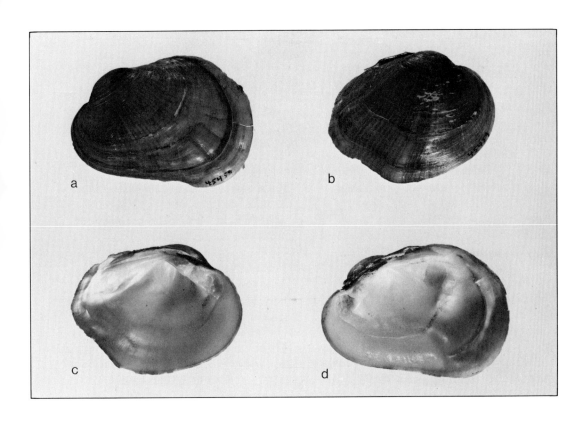

142
Dysnomia torulosa rangiana
a,d: Scioto R., Ohio; 47.6 mm (female).
b,c: Same locality (male).

The female specimen illustrated in colour on page 244 is from Big Darby Creek, Ohio (× 2/3).

143
Dysnomia triquetra
(Rafinesque, 1820)
Tricorn Pearly Mussel

DESCRIPTION
A large male is 55 mm long, 38 mm high, 30 mm wide, and with mid-anterior shell wall almost 6 mm thick. Shell trapezoid-ovate, inflated, with prominent posterior ridge, and with greatest width near the middle. A mature female is 38 mm long, 22 mm high, 27 mm wide, and with mid-anterior shell wall about 3 mm thick. Shell trapezoid, strongly inflated, with bulging posterior ridge, and with greatest width at posterior ridge. Adult females are always smaller than adult males; also in females the highly inflated posterior ridge is extended posteroventrally, thus causing the margin to be expanded, the posterior slope is more strongly flattened than in the male, and both the posterior ridge and slope are marked by strong and wavy radial ribs that denticulate the margin of the shell. In both sexes the periostracum is yellowish, yellowish brown or yellowish green, and covered with numerous broken dark-green rays over the whole shell except that the rays are narrower and partly obscure on the posterior slope. Rays are broken by chevron-shaped pale spots so that the dark spots are also chevron-shaped in many areas. In old specimens the rays disappear near the ventral margin. Nacre whitish. Beaks elevated above hinge line, deeply excavated, and in adults flattened at their apices by mutual abrasion. Beak sculpture composed of 3 or 4 faint bars that are more or less double-looped. Hinge teeth elevated and compressed: pseudocardinal teeth sharp, inclined forward, ragged, 2 in each valve; lateral teeth short, serrated, blade-like, 1 in the right valve and 2 in the left.

Not closely similar to any other species in Canada. Compare with *Truncilla truncata* and *Alasmidonta marginata*.

DISTRIBUTION
Great Lakes system in Lake Erie and in tributaries of Lake Huron, Lake St. Clair, and Lake Erie. Ohio-Mississippi system south to northern Alabama.

ECOLOGY
Characteristic of riffle areas of rivers and creeks as are other species of its genus. Occurs on stony and sandy bottoms and is usually deeply buried. A long-term breeder with gravid period extending from September to May. Glochidia are small, without hooks, and (except for the straight hinge line) nearly circular. The host fish is unknown.

143
Dysnomia triquetra
a,d: Thames R., Chatham, Ont.; 49.2 mm (male).
b,c: Same locality (female).

The female specimen illustrated in colour on page 244 is from the Green R. at Mumfordville, Hart Co., Ky. (× 2/3).

X   Superfamily Sphaeriacea

FAMILY CORBICULIDAE
(Little Basket Clams)

Shell medium-sized to large, bivalved, mostly rather thick, ovate, and with concentric striae. Periostracum yellow, green, or brown. Hinge teeth strong: 3 divergent pseudocardinal teeth; long, ordinarily serrated lateral teeth in front of and behind the pseudocardinals. Foot large. Most species have pelagic larvae. The family is native to North and South America, Africa and Asia, and occurs in brackish water and fresh water.

144
Corbicula fluminea
(Müller, 1774)
Asiatic Clam

DESCRIPTION
(of Washington State specimens)
Shell up to about 50 mm long, high (H/L 0.84-0.91), rather inflated (W/L 0.56-0.67), more or less triangular but variable, and typically with a moderately thick shell. Beaks high and close to centre. Dorsal internal margin open V-shaped, anterior and posterior margins rounded, and ventral margin long and broadly curved. Lateral teeth long, strong, finely serrated, located anterior and posterior to the beaks, and single or double in each valve. Pseudocardinal teeth short, strong, blunt on their tops, and double or triple in each valve. Ligament posterior to beaks, and thick, strong and external. Concentric ribs low but prominent, coarse (about 5 to 10 per cm in centre of shell), and uneven in some specimens. Periostracum thick, deciduous, yellowish brown to blackish brown, and dull to glossy. Young juveniles exhibit about 3 conspicuous brownish-purple radial bands near the umbones. Nacre white, purple, or a combination of white and purple.

Does not closely resemble any other North American species. Immature specimens may be distinguished from *Sphaerium* species by the heavier shell, coarse concentric ribs, roughly triangular shape, and finely serrated hinge teeth; very young specimens are also characterized by radial purplish bands.

North American populations of this species have also been cited as *C. manilensis* (Philippi, 1844) and *C. leana* (Prime, 1864).

DISTRIBUTION
First introduced into the Columbia River system in Washington State about forty years ago, and has since spread south and east throughout most of the southern and central United States. It has not ascended the Columbia River into Canada, nor has it been found alive elsewhere in southern British Columbia or in the Great Lakes, but it is likely to spread into southern Canada in the near future.

ECOLOGY
Sometimes present in incredible numbers. Occurs in lakes, rivers, and canals. Exhibits a preference for substrates of mixed mud and sand. Monoecious and probably capable of self-fertilization. Unlike all other freshwater molluscs in North America the larvae are free-floating veligers; therefore, rapid dispersal is greatly facilitated.

144
Corbicula fluminea
a,d: Columbia R. near John Day Dam, Wash.; 30.2 mm.
b,c: Columbia R. near Carson, Wash.

FAMILY SPHAERIIDAE
(Fingernail Clams and Pea Clams)

Shell small to small-medium, bivalved, thin to slightly thickened, ovate, and with concentric striae. Periostracum yellowish or brownish. Hinge teeth small: 2 very small pseudocardinal teeth in the left valve and 1 in the right; nonserrated lateral teeth before and behind the pseudocardinals, a single set in the left valve and a double set in the right. Foot tongue-shaped. The larvae are held within the mantle cavity and released as crawling young. The family is worldwide in fresh water. Two subfamilies are recognized, Sphaeriinae (Fingernail Clams) and Pisidiinae (Pea or Pill Clams).

SUBFAMILY SPHAERIINAE
(Fingernail Clams)

145
Sphaerium (Sphaerium) corneum (Linnaeus, 1758)
European Fingernail Clam

DESCRIPTION
Shell up to about 9 mm long, relatively high (H/L 0.80–0.88), moderately inflated (W/L 0.52–0.62), ovate, and thin shelled. Beaks broad, very low, and centrally located. Dorsal margin evenly curved and of medium length, ventral margin longer and more flatly curved, anterior margin rounded, and posterior margin rounded or somewhat straight. Hinge plate long, narrow, and evenly curved; hinge teeth compressed. Shell surface covered with fine, evenly spaced concentric striae that are finer near the beaks but obsolete on the beaks. Periostracum brownish and glossy.

Resembles *S. nitidum*, but that species is smaller with a shorter hinge, and the striae are clearly visible on the beaks. Compare also with *S. occidentale* and *S. rhomboideum*.

DISTRIBUTION
Native to Europe and Asia but introduced into the St. Lawrence System. Now occurs in Lake Erie, Lake Ontario, the Ottawa River, the St. Lawrence River, Lake Champlain, and several smaller confluent water bodies in that region.

ECOLOGY
In North America found in large and medium-sized lakes and in slow-moving portions of large and medium-sized rivers. In Europe it occurs in diverse perennial-water habitats. The young are held within each mature individual, and number from 2 to 20 depending on the size and degree of maturity of the adult.

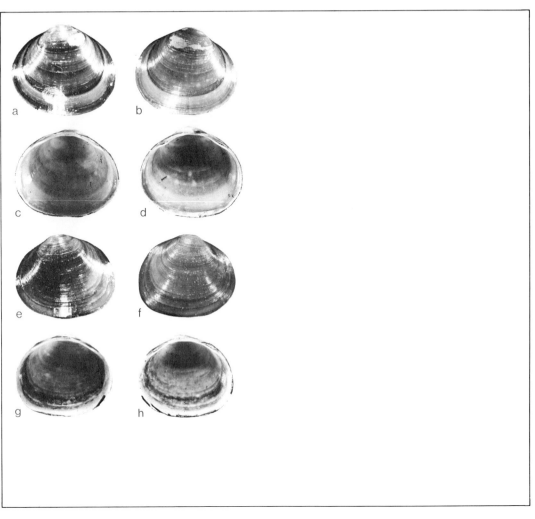

145
Sphaerium corneum
a,b,c,d: L. Ontario near Bath, Ont.; 7.1 mm.
e,f,g,h: Another specimen, same locality; 6.7 mm.

146
Sphaerium (Sphaerium) fabale (Prime, 1851)
River Fingernail Clam

DESCRIPTION
Shell up to about 12 mm long, rather high (H/L 0.80–0.84), compressed (W/L 0.47 0.52), subovate, and of medium thickness. Beaks very low, narrow, and located anterior of centre. Dorsal margin rather long, sharply rounded, and slightly bent anterior of the beaks; ventral margin equally long and more flatly curved; anterior margin gently curved above and more sharply curved below; posterior margin curved sharply above and below and flatly curved centrally. Hinge plate long, almost the full length of the shell, narrow, and bent anterior to the beaks; lateral teeth short, distinct, and distally placed. Surface even and with fine concentric striae on the beaks, but uneven and with coarse and irregularly spaced striae farther out on the shell. Periostracum dull to slightly glossy, and yellowish, yellowish brown or brown.

Resembles the Pacific Coast species *S. patella*, and should be compared with it. Some specimens are similar to *S. striatinum*, but that species is more inflated, and has an even surface and strong striae on the beaks.

DISTRIBUTION
Southern Ontario south to Georgia and Alabama, and west to Michigan.

ECOLOGY
A river and stream species; not recorded from lakes. Substrates inhabited include gravel and also gravelly sand in cracks in a limestone bottom. The life history of this species has not been studied.

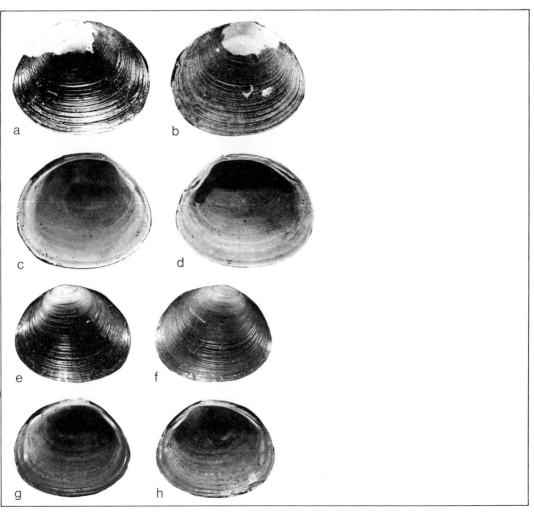

146
Sphaerium fabale
a,b,c,d: Millhaven Creek near Odessa, Ont.; 11.5 mm.
e,f,g,h: Another specimen, same locality; 10.0 mm.

147
Sphaerium (Sphaerium) nitidum Clessin, 1876
Arctic-Alpine Fingernail Clam

DESCRIPTION

Shell up to about 6 mm long, relatively high (H/L 0.80-1.05), somewhat compressed to greatly inflated (W/L 0.58-1.00), ovate to nearly circular, thin shelled, and with low nearly central umbones. Dorsal margin rounded and rather short, ventral margin more openly rounded and much longer, anterior margin round, and posterior margin also rounded but joining the ventral margin with a rounded angle. Hinge plate short and very narrow; lateral cusps all distal, and cardinal teeth slim. Fine concentric striae cover the shell and maintain their height and spacing up over the beak. Well-marked annuli are also present on many specimens. Periostracum shiny and pale yellowish brown.

Resembles *S. occidentale*, but in that species none of the lateral cusps are distal, a radial ridge is present on the inside of each valve, and the habitat is quite different. Compare also with *S. corneum*.

DISTRIBUTION

Labrador and Ungava northwest to Victoria Island and Alaska, and south to the St. Lawrence system; in western mountain lakes south to Utah. Also occurs across northern Eurasia.

ECOLOGY

An arctic and alpine species that thrives in cold water. It occurs in large and small lakes and in rivers of various widths on diverse substrates. Up to 6 young, mostly of varying sizes, have been found in adult specimens. It is a favourite food of arctic fishes.

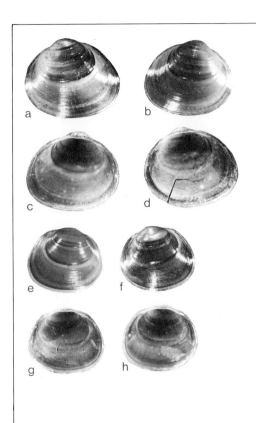

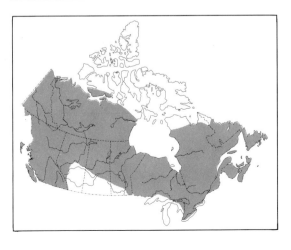

147
Sphaerium nitidum
a,b,c,d: Pigeon Bay, L. Superior, Canada-U.S. boundary; 6.4 mm.
e,f,g,h: Another specimen, same locality; 5.1 mm.

148
Sphaerium (Sphaerium) patella (Gould, 1850)
Rocky Mountain Fingernail Clam

DESCRIPTION
Shell up to about 15 mm long, moderately high (H/L 0.72-0.82), compressed to inflated (W/L 0.47-0.62), more or less ovate, and with shell walls of medium thickness. Beaks low, and slightly anterior. Dorsal margin curved and rather long, ventral margin more openly curved and longer, and anterior and posterior margins obliquely flattened above and rounded below. Hinge plate long and somewhat unevenly curved; lateral teeth short, strong, and at the ends of the hinge plate. Concentric striae very fine on the beak and a little coarser on the body of the shell. Periostracum shiny, yellowish to brownish, and with dark-brown or greenish concentric bands in many specimens.

Similar to the eastern *S. fabale*, but that species is smaller, is dull rather than glossy, and has rather heavy striae on the body of the shell.

DISTRIBUTION
Southern British Columbia to Idaho and northern California in the Pacific drainage.

ECOLOGY
Lakes, sloughs, rivers, and streams. No details of substrate preferences have been recorded. The life history of this species is unknown.

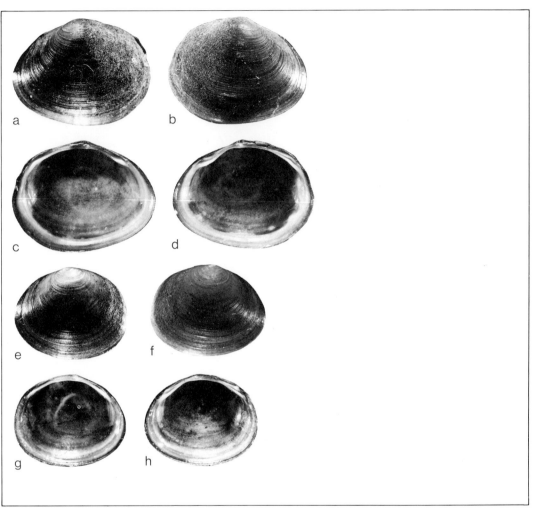

148
Sphaerium patella
a,b,c,d: Abbotsford L., Abbotsford, B.C.; 12.0 mm.
e,f,g,h: Another specimen, same locality; 9.7 mm.

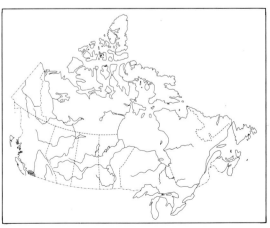

149
Sphaerium (Sphaerium) rhomboideum (Say, 1822)
Rhomboid Fingernail Clam

DESCRIPTION
Shell up to about 14 mm long, moderately high (H/L ca. 0.78-0.88), of medium inflation (W/L 0.48-0.62), more or less rhomboidal in shape, and thin walled but fairly strong. Beaks low, broad, and centrally located. Dorsal margin curved, ventral margin flattened, anterior margin obliquely and flatly curved above and rounded below, and posterior margin obliquely truncate. Hinge plate long, narrow, and unevenly curved. Concentric striae very fine and regular on beaks, and fine on rest of shell. Periostracum glossy and chestnut-brown, but ordinarily with a paler concentric band at the margin and (on some specimens) with one or more earlier pale bands.

Recognized by its more or less rhomboidal shape, fine striae, low beaks, and shining, dark, but pale-banded periostracum. Compare with *S. corneum*, *S. nitidum*, and *S. occidentale*.

DISTRIBUTION
New Brunswick to British Columbia, and Maine to Pennsylvania and Idaho.

ECOLOGY
Occurs in lakes, ponds, and streams. Found in quiet places among vegetation and, most frequently, on muddy bottoms. Adults ordinarily contain approximately 2 to 11 juveniles, which are of different sizes.

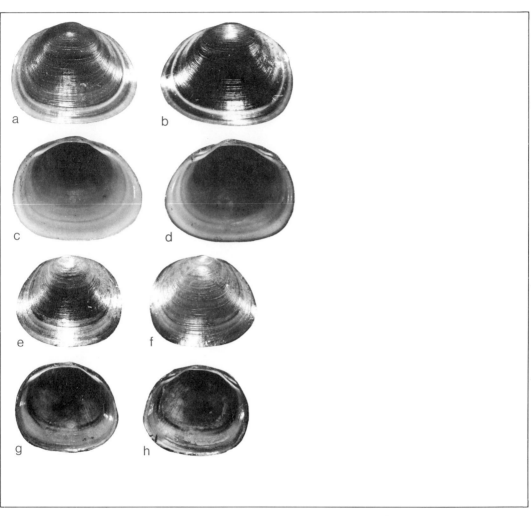

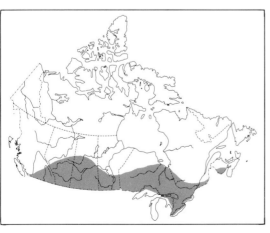

149
Sphaerium rhomboideum
a,b,c,d: Selby Creek, Selby, Ont.; 11.2 mm.
e,f,g,h: Another specimen, same locality; 9.1 mm.

150
Sphaerium (Sphaerium) simile
(Say, 1816)
Grooved Fingernail Clam

DESCRIPTION
Shell up to 25 mm long, of moderate height (H/L 0.72-0.80) and inflation (W/L 0.50-0.58), long-ovate, rather thin to fairly thick, and rather strong. Umbones low and sub-central. Dorsal and ventral margins evenly curved; anterior and posterior margins obliquely flattened above, and roundly pointed below, the centre. Hinge plate narrow; lateral hinge teeth compressed and sharp. Surface of shell covered with heavy, rather coarse, concentric, evenly spaced striae (which number less than 8 per mm in the centre of the shell but are more widely spaced near the beaks). Periostracum brownish or yellowish, but with concentric darker and lighter bands. Shell bluish internally and salmon near the beak cavities.

The large size of the shell and the evenly spaced concentric striae of this species serve to distinguish it from all others.

DISTRIBUTION
Throughout southern Canada, principally in the boreal forest, and south to Virginia, Iowa, and Wyoming.

ECOLOGY
Common. Occurs in all kinds of perennial-water habitats that contain submersed vegetation and muddy or sandy bottoms. A few young of different sizes are present in most adult specimens. The maximum life span of the species has been estimated at about 8 years.

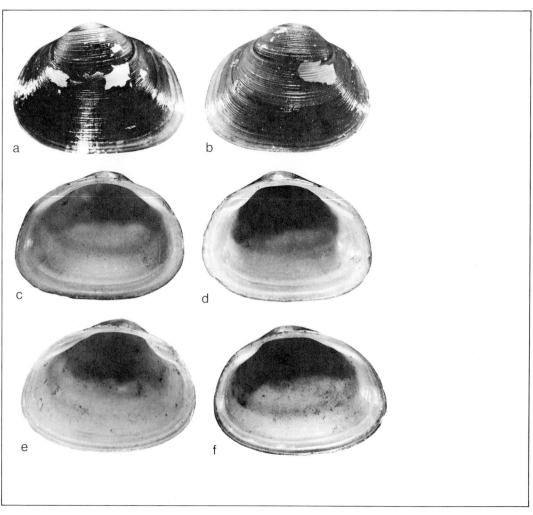

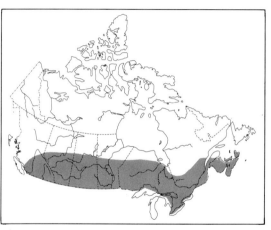

150
Sphaerium simile
a,b,c,d: Wabaskang L., Vermilion Bay, Ont.; 14.6 mm.
e,f: Another specimen but with reversed hinge teeth, same locality; 14.8 mm.

151
Sphaerium (Sphaerium) striatinum (Lamarck, 1818)
Striated Fingernail Clam

DESCRIPTION
Shell up to about 14 mm long, relatively high (H/L 0.78-0.92), moderately inflated (W/L 0.54-0.70), more or less ovate, and rather thick and strong. Umbones low to moderately high and a little anterior of centre. Dorsal and ventral margins evenly curved; anterior and posterior margins more sharply curved or obliquely flattened above the centre. Hinge plate fairly long, of moderate thickness, and unevenly curved; lateral hinge teeth prominent. Surface of shell with concentric, unevenly spaced striae that may be strong or weak on different parts of the shell. Periostracum yellowish to brownish, and with concentric darker and lighter areas in many specimens. Interior bluish to whitish.

Recognized by its medium size, rather inflated and heavy shell, and characteristic concentric striae that are unevenly spaced and strong-to-weak in the same individual but are not weaker on the beaks. Two additional "forms" of *S. striatinum* are recognized, namely form *acuminatum* and form *emarginatum*. For details see Herrington (1962) or Clarke (1973).

DISTRIBUTION
Occurs from New Brunswick to northern Ontario and the Mackenzie River, throughout the United States, and south at least into Mexico.

ECOLOGY
Common. Lives principally in rivers and streams, but also occurs in large lakes and, rarely, in small lakes. Does not occur in swamps, stagnant water, or temporary-water habitats. A few unborn young of different sizes are present in most adults.

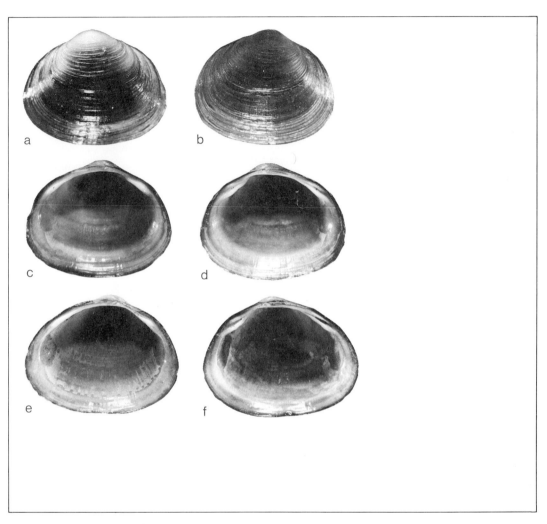

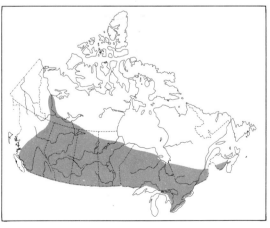

151
Sphaerium striatinum
a,b,c,d: Upper Black R., Pefferlaw, Ont.; 12.3 mm.
e,f: Another specimen, same locality; 13.0 mm.

152
Sphaerium (Herringtonium) occidentale (Prime, 1853)
Herrington's Fingernail Clam

DESCRIPTION
Shell up to about 7 mm long, relatively high (H/L 0.78-0.88), only moderately inflated (W/L 0.45-0.60), evenly ovate, and with thin shell walls. Beaks centrally located, rounded, and elevated. Dorsal margin rounded and of medium length, ventral margin longer and more openly curved, anterior margin rather sharply rounded especially slightly below the centre, and posterior margin more openly rounded. Hinge plate very long, narrow, and with sharp lateral cusps that are near the centre of each tooth or are proximal. Surface covered with fine low concentric striae that are finer near and on the beaks. Within each valve a low but distinct radial ridge occurs, running from the beak cavity to the central ventral margin. Periostracum pale yellowish brown to brown, and dull to somewhat glossy. Adult specimens also exhibit well-marked growth rests.

Somewhat resembles *S. corneum*, but in that species the beaks are only slightly raised above the dorsal margin, not distinctly elevated. The median internal ridge on each valve occurs only in *S. occidentale*. Compare also with *S. nitidum* and *S. partumeium*.

DISTRIBUTION
Discontinuously distributed from Newfoundland and New Brunswick to British Columbia in regions containing calcareous deposits. Its southern distribution extends to Georgia, Utah, and Colorado.

ECOLOGY
Restricted to water bodies that dry up for a part of each year. Typical sites are ditches, swamps, and small shallow ponds, but also occurs among damp leaves. More amphibious than any other North American bivalve mollusc. Adults contain only a few young, about 2 to 5, which are all of the same size.

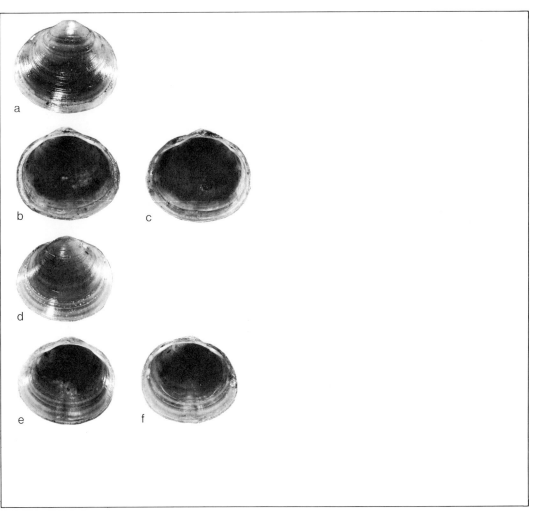

152
Sphaerium occidentale
a,b,c: Pond, Woodstock, Ont.; 7.1 mm.
d,e,f: Another specimen, same locality; 6.6 mm.

153
Sphaerium (Musculium) lacustre (Müller, 1774)
Lake Fingernail Clam

DESCRIPTION

Shell up to about 14 mm long, relatively high (H/L 0.80-0.97), moderately inflated (W/L 0.56-0.67), trapezoidal to rhomboid, and thin walled. Umbones prominent, projecting, normally capped, swollen, and located anterior of centre. Dorsal and ventral margins broadly curved; anterior margin rounded above and a little more sharply rounded below; posterior margin flatly rounded above, sharply rounded below, and longer than the anterior margin. Hinge plate rather long, very narrow, and (in some specimens) almost absent so that the laterals appear to be attached to the shell wall. Lateral teeth slim but distinct. Surface of shell rather glossy and covered with fine to very fine but uneven concentric striae and also (in some specimens) with faint radial lines. Periostracum yellowish brown to brownish.

Two "forms" of *S. lacustre*, in addition to typical *S. lacustre*, are recognized. Form *ryckholti* (Normand, 1844) has higher beaks and a shorter and more rounded dorsal margin. Form *jayense* (Prime, 1851) has a straighter dorsal margin and more truncated anterior and posterior margins. See Herrington (1962) for details.

Resembles *S. partumeium*, but that species has finer and more evenly spaced concentric striae and a more rounded anterior margin, and is glossier. Compare also with *S. securis*.

DISTRIBUTION

Widespread in Canada south of the tree line, throughout the United States (except the southwestern portion), and south into South America. Also in Eurasia, Australia, and the Hawaiian Islands.

ECOLOGY

Occurs in perennial-water lakes, ponds, rivers and streams of all sizes. Ordinarily found on mud but sometimes on sand. Roadside ditches may also be inhabited. A few up to many (1-28) young have been found in Canadian specimens; the young within a single parent are all the same size or of only two different sizes.

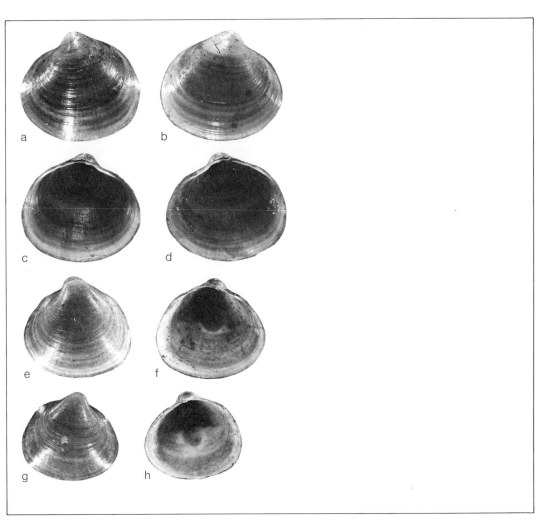

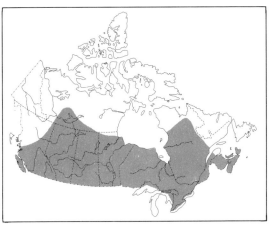

153
Sphaerium lacustre
a,b,c,d: Mocassin L., Denbigh Twp., Ont.; 8.3 mm (typical form).
e,f,g,h: Morph *ryckholti*: pond near Turner Valley, Alta.; *e* and *f* 7.5 mm, *g* and *h* 6.7 mm.

154
Sphaerium (Musculium) partumeium (Say, 1822)
Swamp Fingernail Clam

DESCRIPTION
Shell up to about 7 mm long in Canadian specimens but up to 13 mm in some from farther south, relatively high (H/L 0.84-0.88), of medium inflation (W/L 0.55-0.64), quadrate-ovate, and thin walled. Beaks low, more or less central, and commonly capped. Dorsal margin almost straight and quite long; ventral margin slightly longer than dorsal and well curved; anterior margin rounded; posterior margin truncated and a little rounded, almost perpendicular to the dorsal margin, joining the dorsal margin with a definite but rounded angle, and joining the ventral margin with a rather sharp curve. Hinge plate long, very narrow, and (in some specimens) with lateral teeth apparently arising directly from the shell wall. Lateral teeth narrow but distinct. Concentric striae fine and evenly spaced. Periostracum glossy, smooth, and greyish brown to yellowish brown.

May resemble *S. securis*, but that species is somewhat smaller and has heavier striae, and the anterior ventral margin curves higher, thus causing the anterior margin to be very short. Compare also with *S. lacustre* and *S. occidentale*.

DISTRIBUTION
New Brunswick to Saskatchewan; southeastern British Columbia; and throughout the United States except the extreme southwest.

ECOLOGY
Common. Occurs in large and small lakes, ponds, swamps, vernal ponds and slow-moving streams of all sizes. The usual substrate is mud. A few up to many young (2-30), all in a few discrete size classes, are held within each parent.

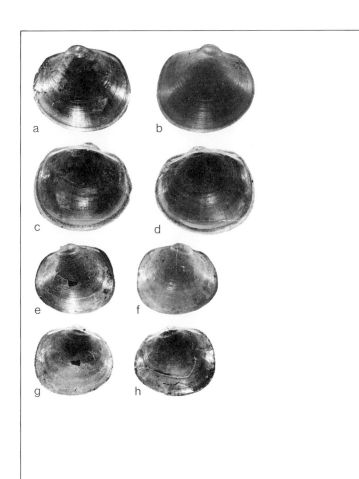

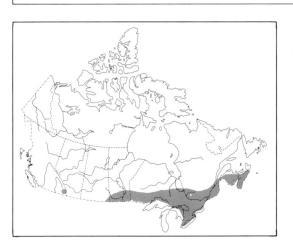

154
Sphaerium partumeium
a,b,c,d: Pond, Sainte-Anne-de-Bellevue, Que.; 6.7 mm.
e,f,g,h: Swamp, Asphodel Twp., Ont.; 5.4 mm.

155
Sphaerium (Musculium) securis
(Prime, 1851)
Pond Fingernail Clam

DESCRIPTION
Shell up to about 6 mm long, relatively high (H/L 0.83-0.92), moderately inflated (W/L 0.54-0.62), quadrate-ovate, and thin walled. Beaks more or less central, somewhat swollen, and capped in many specimens. Dorsal margin rather long and slightly to moderately curved; ventral margin somewhat longer, roundly curved, and swinging high anteriorly, causing a marked shortening of the curved anterior margin; posterior margin roundly truncate, much longer than the anterior margin and forming a 90° angle with the dorsal margin. Hinge plate long (almost the full length of the shell), unevenly curved, very narrow, and bearing slim but distinct hinge teeth. Concentric striae fairly coarse to moderately fine and evenly spaced. Periostracum dull to glossy, yellowish brown to brown, and in many specimens coated with a rusty-brown deposit.

Resembles *S. partumeium*, with which it should be compared.

DISTRIBUTION
Newfoundland and Nova Scotia to British Columbia and southwestern Northwest Territories. Recorded throughout the United States except in the arid southwest.

ECOLOGY
Lives in lakes, ponds, rivers and streams of all sizes, and in both perennial-water and vernal habitats. Mud is the usual substrate, and vegetation is abundant at most occupied sites. A few young in up to three size classes are found in most adult specimens.

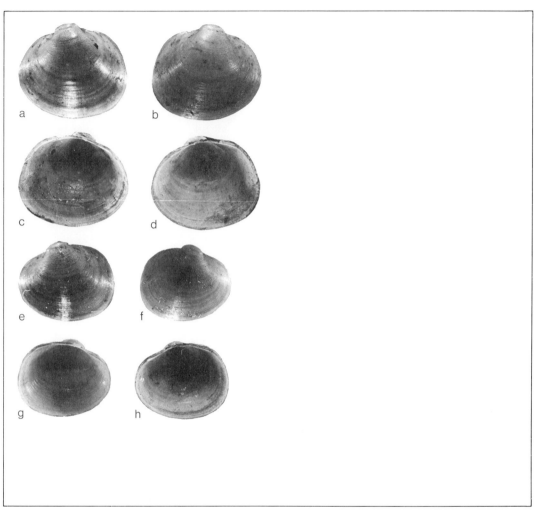

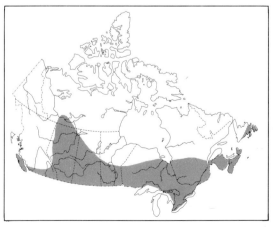

155
Sphaerium securis
a,b,c,d: Temporary pond, Carp, Ont.; 7.5 mm.
e,f,g,h: Another specimen, same locality; 6.4 mm.

156
Sphaerium (Musculium) transversum (Say, 1829)
Long Fingernail-Clam

DESCRIPTION
Shell up to about 15 mm long, relatively low (H/L 0.66-0.77), moderately inflated (W/L 0.47-0.52), elongate-ovate, and thin shelled. Beaks narrow, elevated, capped in some specimens, and anterior of centre. Dorsal margin long and slightly curved; ventral margin even longer and openly curved; anterior and posterior margins flatly curved or obliquely truncate above, sharply rounded below, and (in many specimens) joining dorsal and/or ventral margins at an angle. Hinge plate long, narrow, slightly curved, and bent behind cardinal teeth; lateral teeth compressed and short. Concentric striae very fine on the beaks, and moderately fine and somewhat irregular farther out on the shell. Periostracum yellowish to yellowish brown.

The elongate form of this species makes it quite easy to recognize. Compare with *S. partumeium*.

DISTRIBUTION
Quebec to southwestern Northwest Territories, and British Columbia south to Florida, Texas and Mexico. Absent from the Atlantic Provinces, New England, and most of the far-western United States. Introduced into England.

ECOLOGY
Occurs in lakes, sloughs, and large-to-medium-sized rivers. The usual substrate is mud but may also inhabit sand. A few up to many young (5-34) in up to four size classes have been observed in individual adult Canadian specimens.

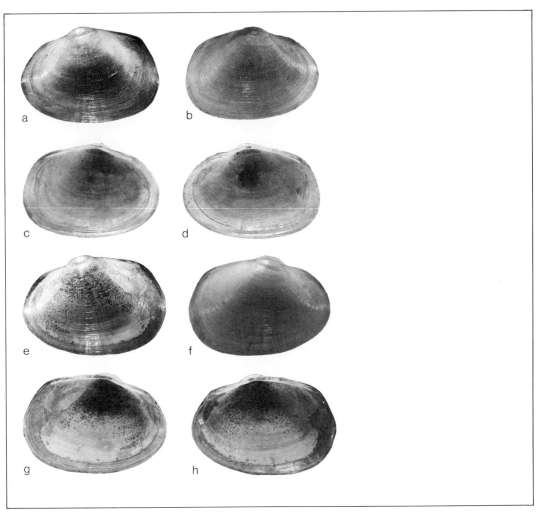

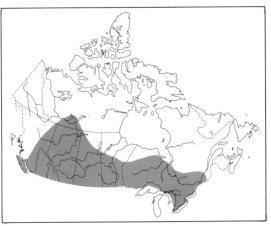

156
Sphaerium transversum
a,b,c,d: L. Winnipeg near Gimli, Man.; 9.3 mm.
e,f,g,h: Another specimen, same locality; 9.8 mm.

SUBFAMILY PISIDIINAE
(Pea Clams or Pill Clams)

157
Pisidium (Pisidium) amnicum
(Müller, 1774)
Greater European Pea Clam

DESCRIPTION

Shell very large for the genus, up to nearly 10 mm long, relatively low (H/L 0.74-0.81), compressed (W/L 0.50-0.60), more or less elliptical, and rather heavy. Beaks low, not inflated, and located about 1/3 of the distance from posterior to anterior. Dorsal margin curved and joining the anterior and posterior margins without angles; anterior margin long, slightly curved above, and rounded centrally; ventral margin long, evenly convex, and passing smoothly into posterior margin; posterior margin flatly curved to subtruncate. Hinge of medium length and rather heavy. Cardinal teeth nearer to the anterior lateral teeth than to the posterior lateral teeth. Striae coarse, far apart (less than 10 per mm), and not becoming obsolete near the beaks. Periostracum yellowish brown and glossy.

Details of tooth structure, according to Herrington (1962), are as follows:

> Cusp of A1 and P1 on distal side of center, of A2 proximal or on proximal side of center, of P2 central or on distal side of center; cardinals fairly close to anterior cusps; C2 resembles a large peg flattened on the anterior side, rounded on the posterior and outside; C4 a thinly curved wedge with thin end out; C3 fits around C2 with a rounded hollow, having a heavy posterior arm reaching to the inner edge of hinge-plate and a lighter anterior arm reaching half-way across hinge-plate; C3 openly and steeply curved, making a good inverted U.

Easily distinguished by its large size, shape, coarse striae that continue up over the beaks, and relative position of the cardinal teeth. Compare with *P. adamsi*, *P. dubium*, and *P. idahoense*.

DISTRIBUTION

Introduced to North America from Europe, where it is widespread. Now lives throughout much of the Great Lakes-St. Lawrence system and in the Delaware River system. It also occurs in North Africa.

ECOLOGY

In North America lives only in large lakes and large rivers. In Europe it is also found in small streams and in canals. Substrates inhabited vary from mud through sand to gravel. Little is recorded about its biology.

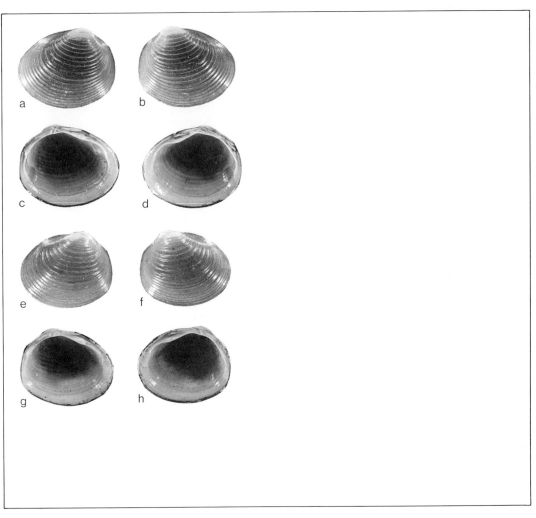

157
Pisidium amnicum
a,b,c,d: Bay of Quinte, L. Ontario, near Adolphustown, Ont.; 6.7 mm.
e,f,g,h: Another specimen, same locality; 6.3 mm.

158
Pisidium (Pisidium) dubium
(Say, 1816)
Greater Eastern Pea Clam

DESCRIPTION

Shell very large for a *Pisidium*, up to about 9 mm long, relatively high (H/L 0.80-0.90), moderately inflated (W/L 0.55-0.63), roughly ovate, and heavy. Beaks broad, prominent, not high, placed at 1/3 of the distance from posterior to anterior, and ridged or flattened in some specimens. Dorsal margin curved, especially posteriorly; anterior margin curved and steep above, rather sharply rounded distally, and curving smoothly into the long, convex ventral margin; posterior margin flatly truncated and vertical. Hinge broad and long. Cardinal teeth somewhat closer to the posterior lateral teeth than to the anterior lateral teeth. Surface striae coarse, concentric, and widely spaced (about 10 to 12 per mm), but fading out on the upper part of the shell. Periostracum yellowish brown or reddish brown, and dull to somewhat glossy.

Details of the hinge teeth, according to Herrington (1962), are as follows:

> Laterals rather short, little more than cusps, and cusps blunt; laterals of right valve divided by a deep pit to receive cusps of left valve; cusps of A1 and P1 on distal side of center, of A2 on distal side of center or distal, of P2 distal; cardinals central or even somewhat toward posterior cusps (this condition has been noticed in no other species). The cardinals are unusual; there is a strong resemblance between those of the left valve and the one of the right valve; the inner ends of C2 and C4, and the posterior end of C3, all reach the inner edge of the hinge-plate; the outer ends of C2 and C4, and the center of C3 reach or almost reach the outer edge of the hinge-plate. . . . C4 is usually a curved wedge. [See Herrington for additional details.]

Resembles *P. amnicum*, with which it should be compared. See also *P. adamsi* and *P. idahoense*. Some variation in tooth shape occurs, for example C2 may be inverted U-shaped or solid, and extending away from C4.

DISTRIBUTION

In Canada known only from southern Ontario in the Lake St. Clair, Lake Ontario and St. Lawrence River drainages. In the United States it ranges from Vermont to Michigan and south to Florida and Alabama.

ECOLOGY

Occurs in rivers and lakes of various sizes. Found on muddy or sandy bottoms. Details of its biology are not known.

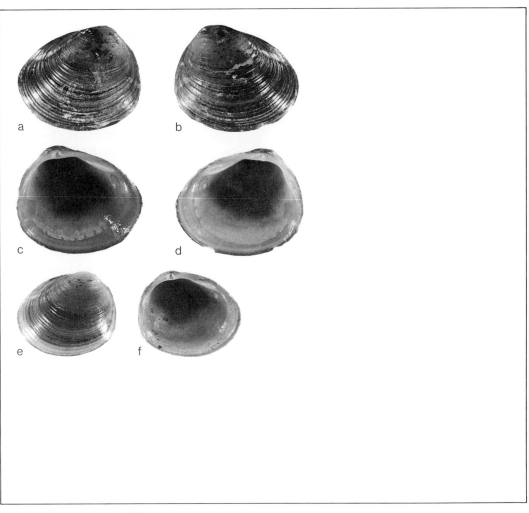

158
Pisidium dubium
a,b,c,d: Ouse R. near McLean, Ont.; 8.6 mm.
e,f: Another specimen, same locality; 6.9 mm.

159
Pisidium (Pisidium) idahoense Roper, 1890
Giant Northern Pea Clam

DESCRIPTION
Shell very large, up to nearly 12 mm long, relatively high (H/L 0.82-0.92), compressed to moderately inflated (W/L 0.47-0.68), roughly ovate, variable in shape, and heavy. Umbones somewhat elevated and inflated, and located about 1/3 of the distance from posterior to anterior. Dorsal margin short and rounded, anterior margin rounded and steep above and roundly pointed below, and ventral and posterior margins openly rounded. Hinge plate moderately heavy and long. Surface covered with fine striae, 15 or more per mm. Periostracum yellowish or yellowish brown, and glossy.

Details of dentition, according to Herrington (1962), are as follows:

> Laterals in large specimens short; cusps rather blunt on top; cusp of A1 on distal side of center or distal, of P1 distal, of A2 proximal or on proximal side of center, of P2 central; cardinals slightly nearer anterior cusps; C2 very short and curved or forming an inverted D; C4 much longer, straight or but slightly curved and directed toward cusp of P2 or a little inside; C3 rather long and considerably curved, about parallel with hinge-plate, center frequently much thinner than ends, posterior end slightly more enlarged than the anterior.

Recognizable by its large size, fine striae, and glossy periostracum. It is the only large *Pisidium* in the Northwest Territories and west of Saskatchewan.

DISTRIBUTION
Occurs from the Great Lakes northwestward to the arctic coast (western Northwest Territories and Alaska), south in the Rocky Mountains to California, and as disjunct populations in Prince Edward Island. Also occurs in Sweden.

ECOLOGY
Most frequently found in cold arctic and mountain lakes, but also lives in small relatively warm lakes in the southern part of its range. A sand bottom with vegetation is its usual habitat. Nothing is known about its life history.

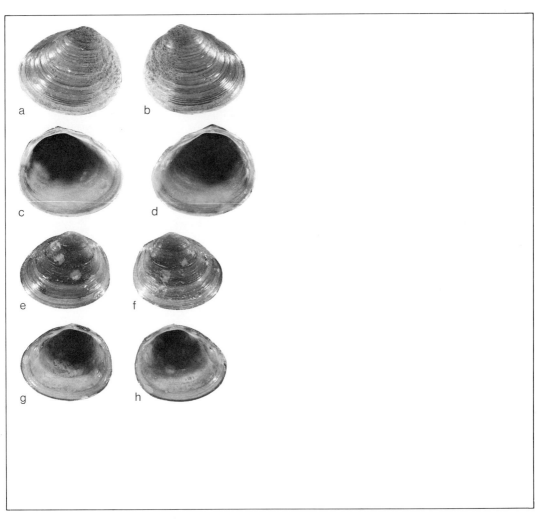

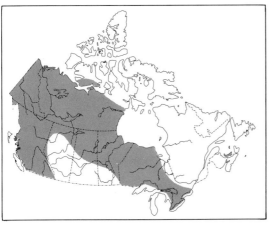

159
Pisidium idahoense
a,b,c,d: McVicar Arm, Great Bear Lake, N.W.T.;
   7.1 mm.
e,f,g,h: Conjuror Bay, Great Bear Lake, N.W.T.;
   6.2 mm.

160
Pisidium (Cyclocalyx) adamsi Prime, 1852
Adams's Pea Clam

DESCRIPTION
Shell large, up to about 7 mm long, quite high (H/L 0.77-0.89), inflated (W/L 0.56-0.70), triangular ovate, and rather heavy. Beaks broad, full, and flattened near their apices (best seen on young specimens). Dorsal margin long, curved in many specimens, and joining the anterior margin with an angle; anterior margin long, curved, steep above, and roundly pointed below; ventral margin long and gently curved; posterior margin truncated and vertical. Hinge plate heavy, long, and curved. Surface sculptured with rather fine, closely spaced concentric striae (more than 15 per mm) and prominent growth rests. Periostracum yellowish brown, somewhat shining, but duller than in most other species.

Hinge teeth details, according to Herrington (1962), are as follows:

> Laterals heavy with heavy cusps, broad on top. Cusp of A1 on the distal side of center; cusps of P1 distal, of A2 proximal, of P2 on distal side of center; cardinals central or on anterior side of center; C3 long, considerably curved, its posterior end a little enlarged; C4 as long as, or longer than C2, slightly curved and usually not directed toward interior of shell, but toward cusp of P2 or just inside of it; C2 varies in length, greatly curved or bent.

Characterized by its large size, flattened apices, relatively fine striae, subtriangular shape, and only moderately shiny periostracum. Compare with the other large species of *Pisidium*, that is *P. amnicum*, *P. dubium*, and *P. idahoense*.

DISTRIBUTION
Found throughout southern Canada from Nova Scotia to northern Ontario and Saskatchewan, and south in the United States to Alabama and Colorado.

ECOLOGY
Lives in lakes, ponds, rivers, and streams of at least 8 m in width. Usually found on muddy bottoms. Litter sizes of from 10 to 66 young have been recorded.

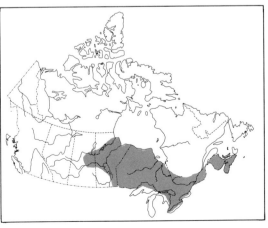

160
Pisidium adamsi
a: Saint John R., Fredericton, N.B. (× 11).
b,c: Another specimen, same locality (× 11).

161
Pisidium (Cyclocalyx) casertanum (Poli, 1795)
Ubiquitous Pea Clam

DESCRIPTION

Shell up to 5 mm long, relatively high (H/L ca. 0.82-0.90), moderately inflated (W/L ca. 0.59 0.77), subovate, highly variable in shape, and of medium thickness. Beaks rounded, somewhat elevated, and located subcentrally or a little posterior of centre. Dorsal and ventral margins gently curved, anterior end moderately long and rounded, and posterior end truncate. Hinge plate more than 3/4 of shell length, and cusps on laterals fairly sharp. Surface of shell finely striate. Periostracum moderately dull to somewhat glossy.

Details of hinge teeth, according to Herrington (1962), are as follows:

Cusp of A1 on distal side of center, of A2 proximal or on proximal side of center, of P2 on distal side of center; P1 and P2 short, cusps distal; cardinals near anterior cusps; C3 slightly curved and somewhat enlarged at posterior end, C2 usually an inverted D, C4 thin and more or less curved, directed toward interior of shell.

Resembles *P. ferrugineum*, but in that species the cusp of A2 has nearly vertical lateral edges instead of edges that are strongly sloped but not vertical. Compare also with other species in the subgenus *Cyclocalyx*.

DISTRIBUTION

Found in all parts of Canada except the Arctic. It is the most widely distributed species of freshwater mollusc in the world, and occurs throughout North and South America, Eurasia, Africa, and Australia.

ECOLOGY

Most abundant Canadian species of *Pisidium*. Lives in lakes, ponds, rivers, small streams, ditches, swamps, and even in temporary-water habitats. Litter size varies from 1 to more than 40. The life span of individuals in a population studied in Michigan is 1 year.

161
Pisidium casertanum
a,b,c,d: L. Villebon near Louvicourt, Que. (*a,b,c* × 11; *d* × 14).

162
Pisidium (Cyclocalyx) compressum Prime, 1852
Ridged-Beak Pea Clam

DESCRIPTION
Shell medium-sized, up to about 5.5 mm long, relatively high (H/L 0.85-0.97), inflated (W/L 0.65-0.75), more or less triangular, and solid. Beaks narrow, prominent, flattened in most specimens, and farther back each typically bears one prominent and concentric ridge. Dorsal margin short and rounded; anterior margin somewhat elongate, gently curved above, and roundly pointed below; ventral margin long and evenly rounded; posterior margin truncate and more or less vertical. Hinge long, heavy, and curved. Surface sculptured with fine concentric striae (15 or more per mm). Periostracum pale yellowish brown, in some specimens with darker concentric bands, and dull to somewhat glossy.

Details of dentition, according to Herrington (1962), are as follows:

> Laterals rather short, being incorporated into the hinge-plate; A3 and P3 tend to curve around the pit of the sulcus, cusps blunt on top; cusp of A1 distal, of P1, A2 and P2 central or on distal side of center; cardinals central; C3 short, considerably curved, posterior end much the larger; C2 short and stout like an inverted D; C4 rather short (but sometimes quite long), slightly curved, and directed toward cusp of P2; space between posterior end of C2 and C4 considerable.

Distinguished by its moderate size, narrow and high beaks that each bear a prominent ridge (in most specimens), and a hinge that exceeds 3/4 of the shell length. Compare with *P. henslowanum* and *P. supinum*.

DISTRIBUTION
Widespread in Canada and Alaska south of the tree line. It also lives throughout the United States and extends into Mexico.

ECOLOGY
Common. Lives in permanent lakes, ponds, rivers, and streams. Found on a variety of substrates. Its usual habitat is among vegetation in shallow water. One litter containing various numbers of young (up to at least 42) is produced each year.

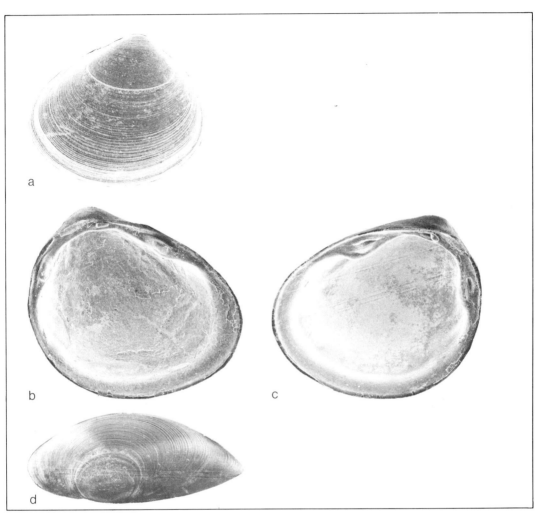

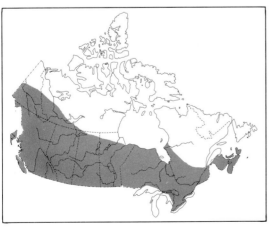

162
Pisidium compressum
a,b,c,d: Baxter Creek near Millbrook, Ont. ($a \times 11$; $b,c,d \times 14$).

163
Pisidium (Cyclocalyx) equilaterale Prime, 1852
Round Pea Clam

DESCRIPTION
Shell up to almost 4 mm long, relatively high (H/L 0.86-0.97) and inflated (W/L 0.60 0.80), nearly circular to ovate, thick shelled, and with large and swollen close-to-centre beaks. Dorsal margin long and rather sharply rounded, ventral margin of medium length and not quite as sharply rounded, anterior and posterior margins even more sharply rounded; all margins together form a nearly circular ellipse. Hinge clearly more than 3/4 shell length, and hinge plate heavy. Surface of shell glossy and covered with moderately fine striae (about 30 per mm).

Hinge details, according to Herrington (1962), are as follows:

> Laterals heavy, rather short, cusps blunt on top; cusps of A1 distal, of P1 on distal side of center or central, of A2 distal or on distal side of center, of P2 central, cardinals central, C3 steeply curved, anterior end rather slim, C2 varies in size, C4 almost straight, directed toward cusp of P2; space between C2 and C4 is a slit directed toward cusp of P2. When a single valve is viewed from the inside, [this species] comes nearest to being round in outline of any of our *Pisidium*.

Best distinguished by its small size, its full, broad beaks that are located subcentrally, its long, heavy hinge, its glossy periostracum, and its circular-elliptical margin.

DISTRIBUTION
Uncommon in Canada. Found from New Brunswick to the vicinity of eastern Lake Superior, and south to Virginia, Pennsylvania, and Illinois.

ECOLOGY
Lives in creeks, rivers and lakes. Typically found on sandy bottoms among vegetation. Known so far only from localities underlain by igneous rocks. No information is available about its anatomy or reproduction.

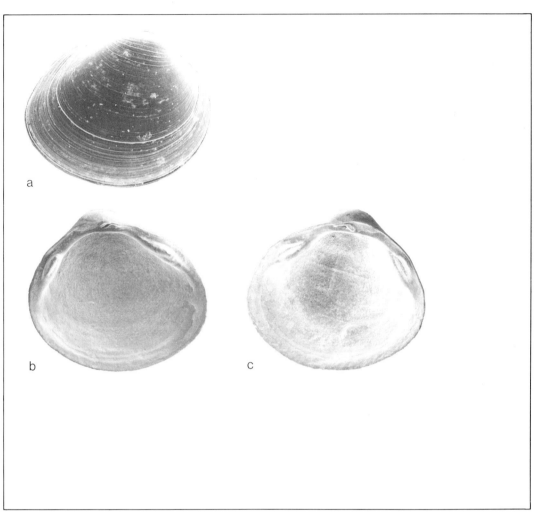

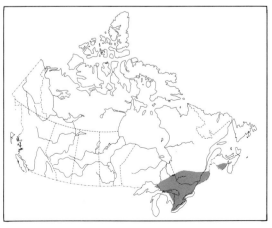

163
Pisidium equilaterale
a,b,c: Baker Brook near Lincoln, N.B. ($a \times 16$; b and $c \times 14$).

164
Pisidium (Cyclocalyx) fallax Sterki, 1890
River Pea Clam

DESCRIPTION
Shell small, up to about 3.5 mm long, moderately high (H/L 0.80-0.90), rather compressed (W/L 0.50-0.62), subovate, and of medium thickness. Beaks a little behind centre and, in some specimens, flattened apically and with a low ridge. Dorsal margin curved; anterior margin flatly curved above and rather sharply convex centrally; and ventral margin curved and passing smoothly into posterior margin that is flatly rounded or roundly truncated. Hinge broad and long. Anterior lateral cusp of the left valve abrupt and slightly twisted counterclockwise toward the interior of the shell. Surface of shell with rather coarse concentric striae (20-30 per mm). Periostracum yellowish brown and dull.

The hinge teeth, according to Herrington (1962), are as follows:

> Laterals not very long; cusps mostly low and not very sharp; cusps of A1 central or on proximal side of center, usually leaning inward (the hinge-plate is much widened here); cusps of P1 central, of P2 central or on proximal side of center, of A2 central or on proximal side of center, not at center of hinge-plate but well inside, which gives it the appearance of a twist so that it is not parallel with the shell margin or directed across it, but directed somewhat inward. This accounts for the great width at A1. A1 and A3 not parallel, but somewhat V-shaped. The cardinals are subcentral to near the anterior cusps; C2 an inverted D; C4 straight or a little curved and directed slightly inside cusps of P2; C3 mostly short, much curved and directed across the hinge-plate, but it varies considerably.

Best recognized by its subovate shape, dull periostracum, and characteristic left anterior lateral hinge tooth.

DISTRIBUTION
Recorded from a number of localities in the southern half of Canada (Quebec to Alberta) and the western Northwest Territories. Also occurs in the northern United States (Maine and New Jersey to Washington) and Alabama.

ECOLOGY
Uncommon. Lives in rivers, streams, and exposed habitats in lakes. Sand or gravel are the usual substrates. Young are born in the spring and, like all other species of *Cyclocalyx*, individuals live for only about 1 year.

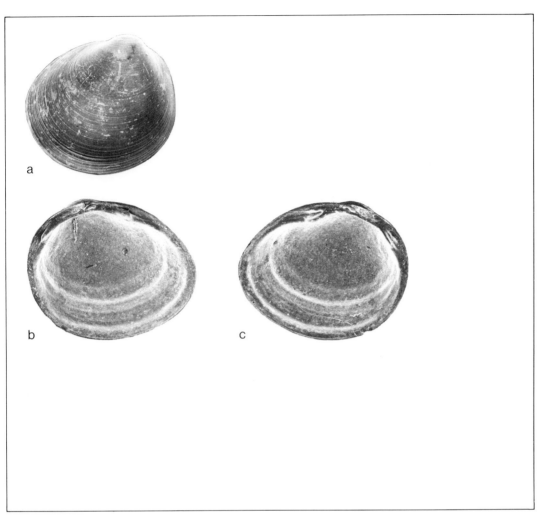

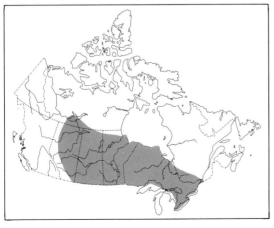

164
Pisidium fallax
a: Bay of Quinte, L. Ontario, near Adolphustown, Ont. (× 11).
b,c: Millhaven Creek near Odessa, Ont. (× 14).

165
Pisidium (Cyclocalyx) ferrugineum Prime, 1852
Rusty Pea Clam

DESCRIPTION

Shell small, up to about 3 mm long, variable in relative height and inflation (H/L 0.55-0.94, W/L 0.63-0.90), ovate to subovate, and thin shelled. Beaks subcentral and tubercular to low and broad. Dorsal and ventral margins equally rounded, anterior margin variable (roundly pointed centrally in many specimens), and posterior margin rounded or subtruncate. Hinge plate narrow and thin. Lateral cusps in the left valve strongly elevated and pointed. Striae coarse to fine, and periostracum glossy but (in many specimens) with a reddish deposit.

The hinge teeth, according to Herrington (1962), are as follows:

> Laterals short with rather pointed cusps; sometimes inner laterals of right valve are curved outward at their outer ends; cusps of A1 central or on distal side of center, of A2 central or on proximal side of center, of P2 fairly distal (cusps of A2 and P2 are short and high with near-vertical ends); cardinals straight or slightly curved, very small and near anterior cusps, particularly in specimens with short dorsal margin; C2 and C4 roughly parallel with each other, but not with the hinge-plate as the space between them more often runs diagonally across it.

This highly variable species is best distinguished by its left anterior and posterior hinge teeth that bear sharp, strongly elevated cusps. Typically the cusp of A2 has nearly vertical lateral edges that are not sloped as in *P. casertanum*. The position of the cardinal teeth near the anterior cusps (not central or subcentral) is also useful in separating it from other common species, such as *P. compressum, P. nitidum*, and *P. variabile*. Compare also with *P. rotundatum*.

DISTRIBUTION

Occurs throughout Canada and Alaska north to a line running from the Ungava Peninsula to southern Victoria Island, and south in the United States to New Jersey, Utah, and Washington.

ECOLOGY

Common. Occurs in lakes, ponds, rivers, and streams. Usually found among vegetation on a sandy or muddy bottom. Little is known about its anatomy or life history.

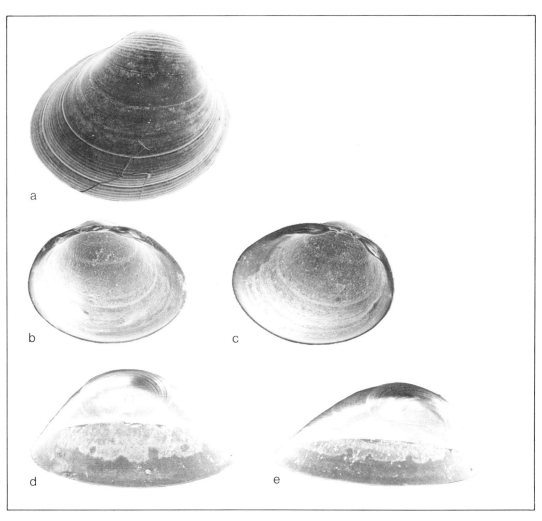

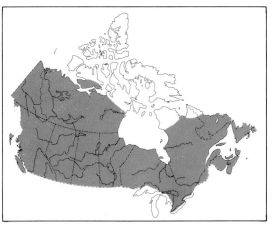

165
Pisidium ferrugineum
a,b,c,d,e: L. Ross, Gatineau Co., Que. (*a,d,e* × 16; *b* and *c* × 14).

166
Pisidium (Cyclocalyx) henslowanum (Sheppard, 1825)
Henslow's Pea Clam

DESCRIPTION
Shell up to about 4 mm long, relatively high (H/L 0.80-0.90), somewhat compressed (W/L 0.45-0.60), triangular-ovate, thin shelled, with anterior end roundly pointed near ventral margin, and (in most specimens) with a prominent roughly concentric ridge on each umbone. Dorsal margin short and curved, joining posterior margin with a rounded angle; posterior margin truncated and flatly curved; ventral margin openly curved; anterior margin roundly pointed below and flatly curved above. Hinge plate less than 3/4 of shell length, gently curved, and tilted backward. Surface of shell dull to somewhat shining and covered with rather coarse concentric striae.

The hinge teeth, according to Herrington (1962), are as follows:

Laterals mostly short; cusp of A1 central to somewhat distal, of P1 distal or on distal side of center, of A2 proximal, of P2 distal; cardinals near anterior cusps; C3 considerably curved, enlarged at posterior end, and almost reaching inner edge of hinge-plate; C4 begins well above C2, angles across hinge-plate, but not as curved downward as in *casertanum*; C2 somewhat as in *casertanum*, posterior end much curved and anterior end begins back from inner edge of hinge-plate.

Well characterized in most specimens by the prominent subconcentric ridge on each beak. In this it resembles the more abundant and widespread *P. compressum*, but in that species the hinge plate is more than 3/4 the length of the shell, the shell is proportionately much higher, and the anterior end is not extended and pointed basally. Compare also with *P. lilljeborgi* morph *cristatum* and *P. subtruncatum*.

DISTRIBUTION
Introduced, probably from Europe, and now widespread in the lower Great Lakes-St. Lawrence River system. Also occurs in Eurasia and Iceland.

ECOLOGY
Rare. In North America occurs chiefly in large bodies of water. In Europe it is also found in unpolluted rivers and canals. Litter sizes of from 1 to 7 young have been recorded in Sweden.

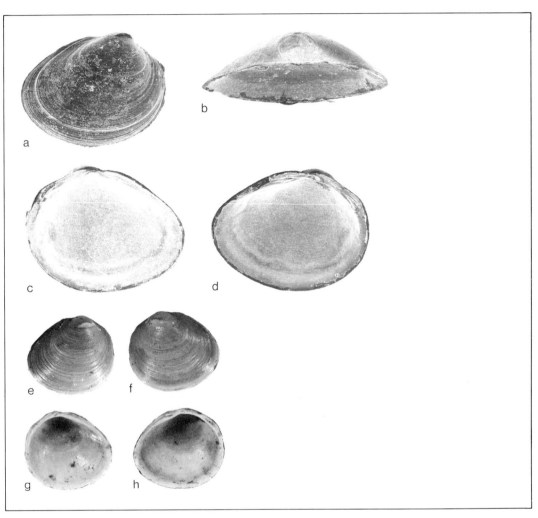

166
Pisidium henslowanum
a: L. Ontario near Collins Bay, Ont. (× 16).
b,c,d: Prince Edward Bay, L. Ontario, Ont. (*b* × 22; *c* and *d* × 19).
e,f,g,h: L. Ontario near Collins Bay, Ont. (× 11).

167
Pisidium (Cyclocalyx) lilljeborgi Clessin, 1886
Lilljeborg's Pea Clam

DESCRIPTION

Shell up to about 4 mm long, proportionately high (H/L 0.82-0.94), moderately inflated (W/L 0.59-0.73), triangular-ovate, and rather thin shelled. Beaks prominent, high, full, and located posterior of centre. Dorsal margin short, posterior of centre, openly curved and joining anterior and posterior margins with angles; posterior margin roundly truncate and vertical; anterior margin elongate and distally rather sharply rounded; ventral margin long and curved. Hinge plate less than 3/4 of shell length and rather heavy. Surface of shell finely to rather coarsely striate, and slightly dull to moderately glossy.

Details of hinge teeth, according to Herrington (1962), are as follows:

Laterals A1, A2 and P2 moderately long (A1 and A3 form a V); cusp of A1 proximal or on proximal side of center, or central (somewhat blunt on top), of A2 proximal (high and sharp on top), of P2 distal (high, a rather steep incline and sharp on top); cardinals close to anterior cusps; C3 rather long, usually bent with anterior end parallel with hinge-plate, posterior end quite expanded; C2 short and broad; C4 narrow, twice or more as long as C2 [but only as long as C2 in some specimens] . . . and extending . . . diagonally along hinge-plate.

Occasionally specimens have ridges on the beaks and may be shorter than the typical form. These are called *P. lilljeborgi* morph *cristatum* Sterki, 1928.

Differs from *P. casertanum* in that the anterior margin joins the dorsal margin at an angle, and the beaks are more elevated and more prominent. Compare also with *P. subtruncatum*.

DISTRIBUTION

Occurs throughout Canada as far north as southern Baffin Island and southern Victoria Island, throughout Alaska, in New England and the northern tier of states, and south in the Rocky Mountains to Colorado, Utah, and California. Also found in Iceland and in northern Europe.

ECOLOGY

Common. Lives in all permanent-water habitats, especially lakes. Inhabits clay, mud, sand, or gravel. Adult specimens containing mature young have been found only in summer, and litter sizes of up to 13 have been observed.

NOTE

Another species of *Pisidium* (*P. waldeni* Kuiper, 1975), closely related to *P. lilljeborgi*, has recently been recorded from arctic Canada. It is reported to differ by having a longer dorsal margin, a longer hinge plate, and shorter cardinal teeth in the left valve, as well as other characters. See Kuiper (1975) for details.

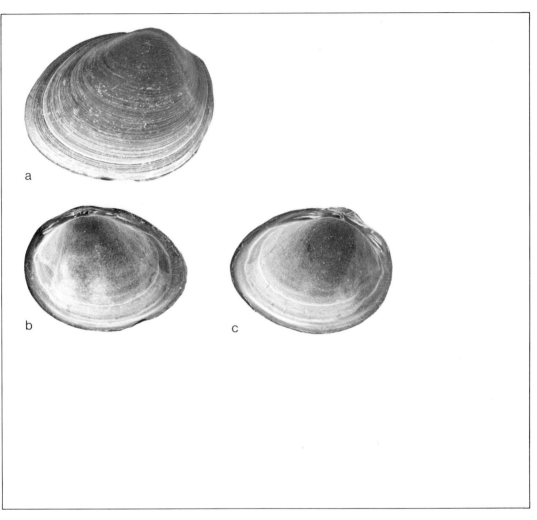

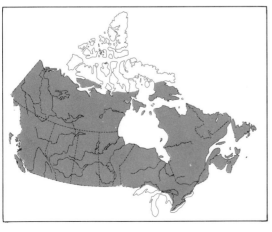

167
Pisidium lilljeborgi
a: Otter L. near Shawville, Que. (× 14).
b,c: L. Saint-Simon, Rimouski Co., Que. (× 11).

168
Pisidium (Cyclocalyx) milium
Held, 1836
Quadrangular Pill Clam

DESCRIPTION

Shell small, up to 3 mm long, of moderate height (H/L 0.62-0.88), greatly inflated (W/L 0.60-0.95), and variable in shape but more or less triangular and flattened ventrally (in end view). Beaks swollen, prominent, and located rather far back. Dorsal and ventral margins openly curved; posterior margin truncated; and anterior margin long, sloped above, and roundly pointed below. Hinge plate narrow and less than 3/4 the shell length. Surface with rather fine concentric striae and a few prominent growth rests. Periostracum thin, glossy, and pale yellowish brown.

The dentition, according to Herrington (1962), is as follows:

> Laterals rather short; cusps inclined to be sharp on top; cusps of left valve toothpick-like, of A1 and A2 central or on distal side of center, of P2 somewhat distal; cardinals fairly near anterior cusps, but varying, sometimes subcentral; C3 slightly curved and uniform in width; C2 and C4 nearly same thickness and about parallel; C4 begins well over C2, slightly curving or straight (C2 is the shorter of the two).

The small size, roughly triangular shape, prominent inflation, and flat but truncated ventral margin formed by the two appressed valves (when viewed from one end) are good diagnostic characters for this species. It resembles *P. nitidum*, but that species is higher, has a more sharply curved ventral margin, and the anterior cusps are more distal and not as sharp.

DISTRIBUTION

Occurs from the Maritime Provinces to British Columbia; in western Canada to north of Great Slave Lake. Also found in Alaska, the northern United States, south in the Rocky Mountains to Colorado and Utah, and in Europe.

ECOLOGY

Uncommon. Lives in lakes, ponds, and slow-moving streams. Occurs on muddy bottoms among vegetation. Swedish specimens were observed to produce young from June to September.

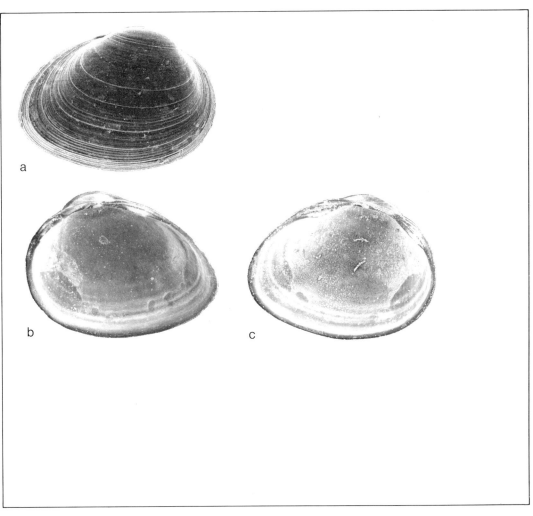

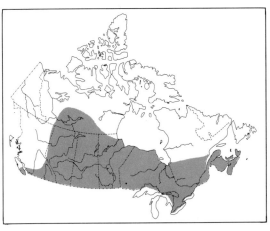

168
Pisidium milium
a,b,c: Creek, Otonabee Twp., Ont. ($a \times 19$; b and $c \times 16$).

169
Pisidium (Cyclocalyx) nitidum Jenyns, 1832
Shiny Pea Clam

DESCRIPTION
Shell small, up to about 3 mm long, relatively high (H/L 0.80-0.92), of moderate inflation (W/L 0.57-0.64), roughly rhomboidal in shape, and thin walled. Beaks posterior of centre, broad, low, and curved. Dorsal margin long and evenly curved; anterior margin curved and steep above and roundly pointed below; ventral margin long and broadly curved; and posterior margin roundly truncated and vertical or undercut. Surface sculptured with fine striae (more than 30 per mm). Periostracum pale yellowish or greyish brown and very glossy.

The hinge teeth, according to Herrington (1962), are as follows:

> Laterals of moderate length, straight or flaring outward at distal end; cusps rather prominent, but inclined to be blunt on top; cusp of A1 distal or on distal side of center, of P1, P2, and A2 rather distal; cardinals subcentral; C3 gently curved, of uniform width except at posterior end, and almost parallel with hinge-plate; C2 slightly heavier than C4; C4 straight or gently curved, about parallel with C2, space between the two of uniform width, straight or a little curved, and usually directed across hinge-plate at a gentle angle.

Distinguished by its small size, relatively long hinge, shiny finely striate surface, and details of dentition. Two well-marked "forms" are known in addition to the typical form. Form *contortum* is longer, has a nearly straight ventral margin, and is more pointed in the anterior basal region. Form *pauperculum* is shorter, higher, heavier, and has a more curved dorsal margin and more-central cardinal teeth.

DISTRIBUTION
Broadly distributed throughout Canada, the United States, Mexico, Eurasia, and North Africa.

ECOLOGY
Common. Lives in all kinds of perennial-water habitats, on various substrates, and most commonly in shallow water. In each litter, 2 to 7 young are ordinarily produced.

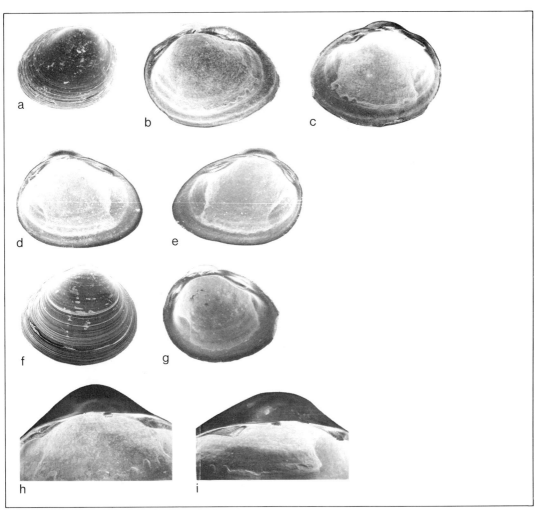

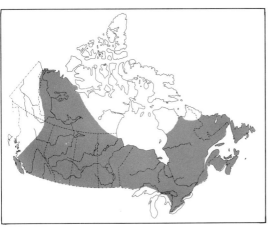

169
Pisidium nitidum
a,b,c,h,i: Typical *nitidum*: Pleasant Park, Cressy, Ont. (*a,b,c* × 11; *h* and *i* × 18).
d,e: Morph *contortum*: Klotz L. near Longlac, Ont. (× 11).
f,g: Morph *pauperculum*: Rice L., Peterborough Co., Ont. (× 11).

170
Pisidium (Cyclocalyx) rotundatum Prime, 1852
Fat Pea Clam

DESCRIPTION
Shell small, up to about 3.3 mm long, relatively high (H/L 0.80-0.92), greatly inflated (W/L 0.70-0.76), ovate, and thin. Beaks prominent, large, inflated, capped in some specimens, and located about 1/3 of the distance from posterior to anterior or (usually) more centrally. All margins rounded, the anterior and posterior margins more sharply rounded than the dorsal and ventral. Hinge plate short, narrow, curved, and posterior of centre. Surface covered with moderate-to-fine striae that are evenly spaced. Periostracum yellowish brown and glossy.

The hinge teeth, according to Herrington (1962), are as follows:

> Laterals short; cusps short and high with near-vertical ends; cusps of A2 proximal, of P2 and A1 central or on distal side of center; cardinals close to anterior cusps; C3 curved, but not much enlarged at posterior end. . .; C2 and C4 short; C2 almost parallel with hinge-plate, straight, sometimes slightly curved, or just a peg; C4 straight or slightly curved, sometimes parallel with hinge-plate, but more often directed slightly downward, then not parallel with C2; proximal end of posterior sulcus of right valve closed by a pseudocallus on inner side of proximal end of P3 and, therefore, does not run out on top of hinge-plate.

Distinguished from its close relative *P. ventricosum* by the more central position of the beaks and the narrow hinge plate (between the cardinals and A2). In *P. ventricosum* the beaks are nearly posterior, and the hinge plate is broad between the cardinals and A2.

DISTRIBUTION
Widespread in Canada south of the tree line. Also occurs in the northern tier of states in the United States, and south in the Rocky Mountains to Mexico.

ECOLOGY
Lives, as does *P. ventricosum*, in lakes, permanent ponds, rivers, and streams. Found among vegetation and in various substrates but typically in mud. Nothing is known about its life history.

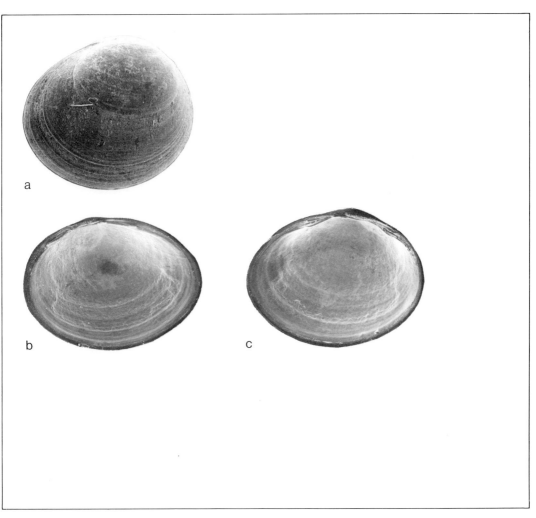

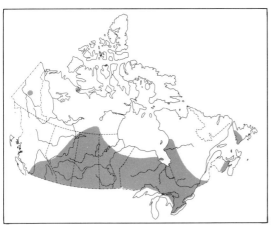

170
Pisidium rotundatum
a: Pond in Byron Bog, London, Ont. (× 23).
b,c: Keefer L. near Kelowna, B.C. (× 19).

171
Pisidium (Cyclocalyx) subtruncatum Malm, 1855
Short-ended Pea Clam

DESCRIPTION

Shell rather small, up to almost 4 mm long, relatively high (H/L 0.74-0.94), inflated (W/L 0.54 0.74), ovate, variable in shape, and thin walled. Beaks narrow, prominent, projecting above the hinge line, and located far back. Dorsal margin short, curved, posterior of centre and joining anterior margin with an angle close to the anterior cusps; anterior margin long or short and rather sharply rounded centrally; ventral margin long and openly curved; posterior margin roundly truncate and joining dorsal margin with an angle. Hinge short and curved; hinge plate narrow. Striae evenly spaced, fine, and numbering 30 or more per mm. Periostracum glossy and yellowish brown.

The hinge teeth, according to Herrington (1962), are as follows:

Laterals rather long, cusps prominent; cusps of A1 and A2 proximal to central, of P1 and P2 central; cardinals near anterior cusps or subcentral; C2 and C4 roughly parallel, posterior ends slightly nearer inner edge of hinge-plate; C3 long and not much curved; width of hinge-plate and size of shell influences cardinals considerably.

Resembles *P. walkeri*, but that species is more compressed, the cusp of A1 is less proximal, the left cardinals are not parallel, and the periostracum is dull. Compare also with *P. lilljeborgi*.

DISTRIBUTION

Occurs in Prince Edward Island and throughout central Canada north to, and slightly beyond, the tree line. Also found from New York to Montana, south in the Rocky Mountains to California and Colorado, and in Europe.

ECOLOGY

Lives in lakes, ponds, muskeg pools, rivers, and streams. Found among vegetation and on various kinds of bottoms. Two litters of young are born each year, and life spans exceed 1 year.

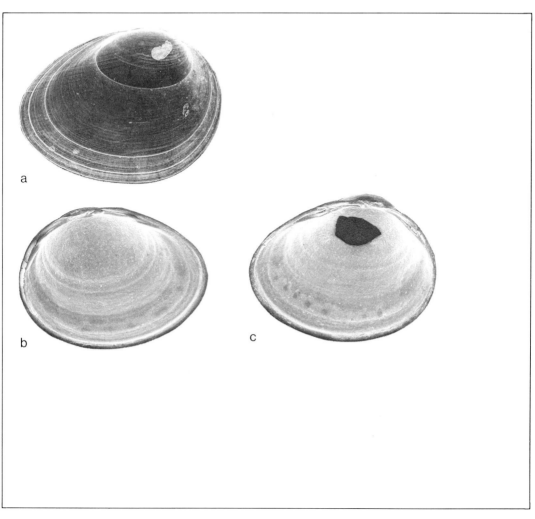

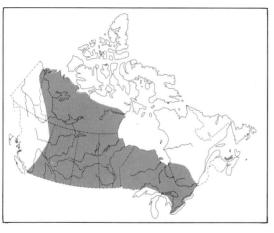

171
Pisidium subtruncatum
a,b,c: Mills L., near Great Slave L., N.W.T. (× 16).

172
Pisidium (Cyclocalyx) supinum Schmidt, 1850
Hump-backed Pea Clam

DESCRIPTION
Shell medium-sized, up to about 4.5 mm long, relatively high (H/L ca. 0.90), inflated (W/L ca. 0.67), roughly trianglular, and moderately thick to thick. Beaks high and full; and each beak surmounted by an oblique, more or less concentric raised ridge. Dorsal margin short, strongly curved, and joining the anterior and posterior margins without angles; anterior margin lengthened, flattened above, roundly pointed below, and curving smoothly into the long rounded anterior margin; posterior margin truncated and only slightly rounded. Hinge teeth typically thick and heavy. Surface sculptured with moderately heavy, concentric striae (about 16 to 18 per mm). Periostracum yellowish brown and moderately dull.

Hinge teeth similar to those of *P. henslowanum*, but much heavier and with the cardinals farther from the anterior cusps. Compare also with *P. compressum* and *P. subtruncatum*.

DISTRIBUTION
This Eurasian and Icelandic species was first discovered in North America near the eastern end of Lake Ontario in about 1959. It has not yet been observed elsewhere in the Great Lakes-St. Lawrence system. Living specimens were recently (1975) found, however, in the Eastmain and La Grande rivers in northern Quebec by B. T. Kidd. Has also been recorded in late Pliocene and Early Pleistocene deposits in Idaho.

ECOLOGY
Uncommon in North America. In Europe it is characteristic of rivers and is rare in lakes.

172
Pisidium supinum
a,b,c: Athol Bay, L. Ontario, near Athol, Ont. (× 14).

173
Pisidium (Cyclocalyx) variabile Prime, 1852
Triangular Pea Clam

DESCRIPTION
Shell up to about 5 mm long, proportionately high (H/L 0.72-0.97), of medium inflation (W/L 0.54-0.68), triangular-ovate, thick walled, and variable in shape. Beaks prominent, full, and located posterior of centre. Dorsal and posterior margins round and continuous; ventral margin long and more openly curved; anterior margin roundly angled distally and joining the short dorsal margin without an angle. Hinge plate about 3/4 the length of the shell (or a little more), heavy, and rather steeply curved. Surface of shell glossy, and covered with rather fine to quite coarse striae (but less than 30 per mm).

The hinge teeth are described by Herrington (1962) as follows:

Laterals rather short, incorporated into hinge-plate; A3 and P3 tend to curve around pit of sulcus; cusps blunt on top; cusps of A1 distal, of P1, A2, and P2 central or on distal side of center; cardinals central; C3 short, much curved, posterior end much the larger; C2 short and stout like an inverted D; C4 fairly short, only slightly curved and directed toward cusp of P2; considerable space between posterior ends of C2 and C4.

Distinguished by its rather heavy and roughly triangular shell, glossy surface, and characteristic hinge teeth. The hinge teeth are similar only to those of *P. compressum*, but that species is smaller, with dull periostracum, and in most specimens bears a prominent concentric ridge on each beak.

DISTRIBUTION
Occurs throughout southern Canada and northward into the subarctic. Also recorded from most of the United States.

ECOLOGY
Common. Lives in virtually all natural perennial-water habitats. Found in various substrates (most frequently mud) and usually amid vegetation. Litter size varies from at least 12 to 34. No anatomical studies have been done on this species.

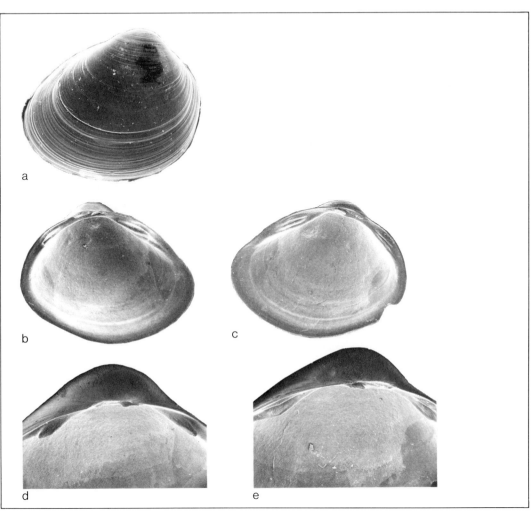

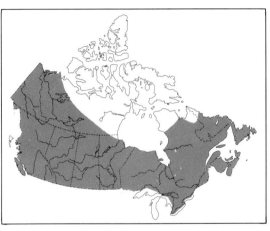

173
Pisidium variabile
a,b,c,d,e: Pond near Perth, N.B. (*a,b,c* × 7; *d* and *e* × 11).

174
Pisidium (Cyclocalyx) ventricosum Prime, 1851
Globular Pea Clam

DESCRIPTION

Shell small, up to about 3 mm long, relatively high (H/L 0.82-1.00), greatly inflated (W/L 0.80-0.95), ovate, and rather heavy. Beaks large, swollen, and placed far back on the shell. Dorsal margin smoothly curved and passing imperceptibly into anterior margin; anterior margin sloped, broadly curved above, and roundly curved centrally or in an area just below the centre; ventral margin long, evenly rounded, and continuing in an uninterrupted curve into the posterior margin; posterior margin flatly curved and undercut. Hinge plate heavy and relatively short, that is less than 3/4 the length of the shell. Surface finely striate (more than 30 per mm) and with prominent growth rests in many specimens. Periostracum yellowish brown to greyish, and glossy.

Hinge teeth similar to those in *P. rotundatum* but heavier, and the hinge plate between the cardinal teeth and A2 is relatively wide, not thin.

This small species is well characterized by its greatly inflated heavy shell and prominent, swollen, posteriorly-placed beaks. Compare with *P. rotundatum* and *P. milium*.

DISTRIBUTION

Occurs throughout central Canada south of the tree line and in isolated localities in eastern Canada. Also recorded from across the northern United States from Maine to Washington, and south in the Rocky Mountains to Mexico.

ECOLOGY

Lives in perennial-water lakes, ponds, rivers and streams of all sizes. Aquatic vegetation and muddy bottoms characterize typical habitats. Nothing is known about its anatomy or reproduction.

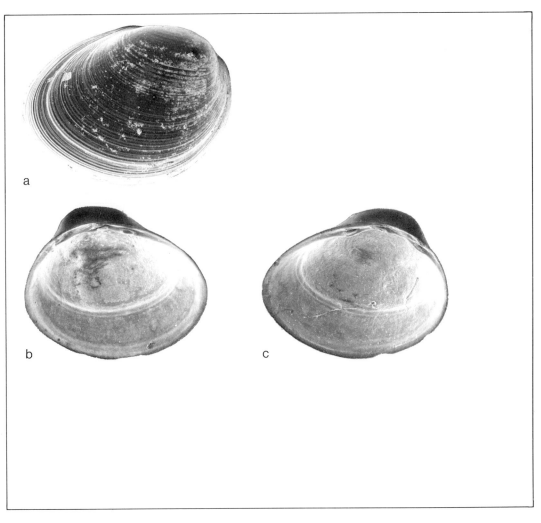

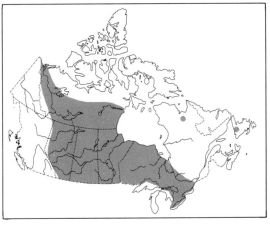

174
Pisidium ventricosum
a: Mechanic Lake Brook near Alma, N.B. (× 27).
b,c: Reindeer L., Brochet, Man. (× 19).

175
Pisidium (Cyclocalyx) walkeri
Sterki, 1895
Walker's Pea Clam

DESCRIPTION

Shell up to about 6 mm long, relatively high (H/L 0.80-0.90), moderately inflated (W/L 0.54-0.68), somewhat ovate but anteriorly elongate, and rather thin walled. Umbones moderately large and located posterodorsally. Dorsal margin strongly curved and tilted posteriorly; anterior margin long, slightly curved above and roundly pointed below the centre; ventral margin long and gently curved; posterior margin flattened and vertical. Hinge plate less than 3/4 the shell length and not entirely parallel with the dorsal margin. Surface of shell dull to somewhat shining and with less than 30 striae per mm.

The hinge teeth, according to Herrington (1962), are as follows:

> Laterals rather long, cusps moderately sharp; cusps of A1 central or on distal side of center; of P1 distal, of A2 proximal or on proximal side of center, of P2 central; cardinals subcentral; C2 short, much curved (sometimes an inverted D); C4 much lighter, short, curved, directed toward interior of shell; C3 parallel with hinge-plate, varying in degree of curvature, its posterior end enlarged.

The "form" of *P. walkeri* called *mainense* Sterki, 1903, is smaller and relatively shorter, and with finer striae, a more rounded anterior slope, left cardinal teeth that are more nearly parallel, and a longer C2.

Differentiated from *P. subtruncatum* by its coarser striae. Compare also with *P. casertanum*.

DISTRIBUTION

Typical *P. walkeri* have been found from New Brunswick to James Bay, and in western Canada north into the Northwest Territories. Also widely distributed in the United States south to Virginia and Arizona. *P. walkeri* form *mainense* is more eastern, and is recorded from scattered sites in subarctic eastern Canada, from New Brunswick to Massachusetts, and in Illinois. Both forms are considered together on Map 175.

ECOLOGY

Rather uncommon. Occurs in permanent lakes, ponds, rivers, and streams. Found most frequently among vegetation and on a variety of substrates. A single litter of young are born during the 1-year lifetime of each individual. The anatomy of this species has not been studied.

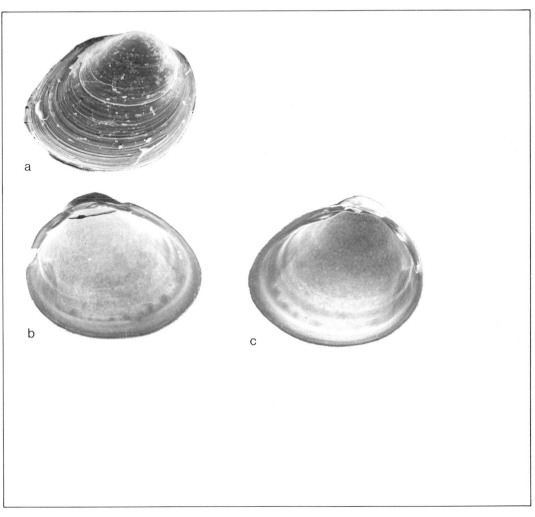

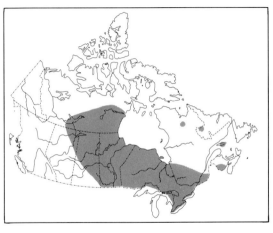

175
Pisidium walkeri
a,b,c: L. Athabasca, "The Willows", Alta. ($a \times 16$; b and $c \times 14$).

176
Pisidium (Neopisidium) conventus Clessin, 1877
Arctic-Alpine Pea Clam

DESCRIPTION
Shell up to almost 3 mm long, of moderate height (H/L 0.68-0.88), rather compressed (W/L 0.44-0.65), more or less ovate or trapezoid, variable in shape, thin, and fragile. Beaks low, rounded, and located a little posterior of centre. All margins curved and with variable shape: for example, in some specimens the anterior and posterior margins are obliquely truncated and roughly parallel, in others the anterior margin is roundly pointed centrally and the posterior margin is vertically truncated. Hinge plate very long and narrow. Surface of shell covered with fine concentric striae and irregularly spaced growth rests. Periostracum thin and pale yellow or whitish.

Details of the hinge teeth, according to Herrington (1962), are as follows:

Laterals long and slender; cusps of A1, P1, and A2 distal or on distal side of center, of P2 distal; cardinals central or subcentral; C2 at or on proximal end of A2, short, close to inner edge of hinge-plate or overhanging and either about parallel with hinge-plate or with posterior end slightly more interior; C4 slightly longer than C2, slimmer, straight or slightly curved, beginning above center of C2 and parallel with it, or its posterior end directed a little more toward the interior; C3 moderately long, slightly curved, almost parallel with hinge-plate, posterior end enlarged somewhat, therefore, nearer inner edge of hinge-plate.

Well characterized by its small, thin, fragile, and pale-coloured shell, its overhanging cardinal hinge teeth, its subtrapezoidal shape (when present), and its unusual habitat.

DISTRIBUTION
Occurs in most of the nothern states of the United States and northward throughout most of Canada (to Victoria Island) and Alaska. Also found in alpine lakes across Eurasia and north to the Eurasian arctic coast.

ECOLOGY
A cold water species. Lives principally at considerable depths in large lakes within the temperate parts of its range, and at all depths within subarctic and arctic regions. Two broods of young are produced each year, 1 in summer and 1 in winter.

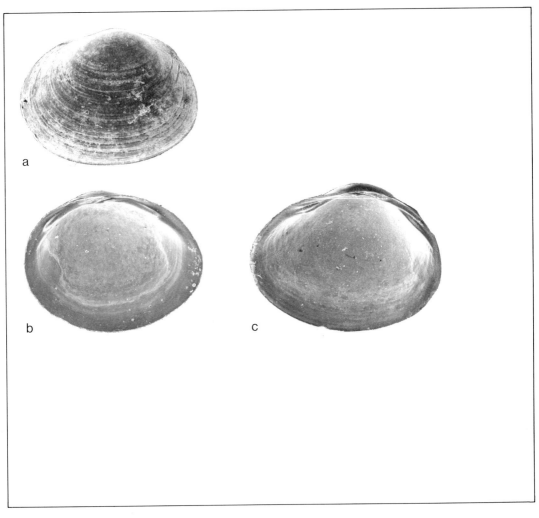

176
Pisidium conventus
a,b,c: Crow L., Frontenac Co., Ont. (*a* and *c* × 19; *b* × 27).

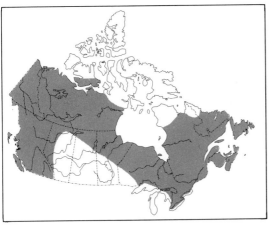

177
Pisidium (Neopisidium) cruciatum Sterki, 1895
Ornamented Pea Clam

DESCRIPTION
Shell up to about 2 mm long, proportionately high (H/L ca. 0.94-0.95), quite inflated (W/L ca. 0.70-0.74), triangular, thick, and heavy. Beaks prominent, elevated, and posterior of centre. Each beak bears a heavy U-shaped ridge whose ends are directed ventrally and perpendicular to the lines of growth. Dorsal margin short and rounded, ventral margin long and flatly curved, anterior margin steep and roundly pointed basally, and posterior margin truncated and joining the dorsal margin in a smooth curve. Hinge plate very heavy and more than 3/4 the length of the shell. Surface dull and with coarse concentric striae (about 16 per mm).

Details of the hinge teeth, according to Herrington (1962), are as follows:

Laterals heavy and short, little more than cusps; cusps of A1 and P1 distal, of A2 central or on either proximal or distal side of center, of P2 on distal side of center; cardinals central; C2 heavy, usually an inverted D; C4 slim, slightly curved and directed toward cusp of P2 or slightly inside; C3 parallel with hingeplate, curved somewhat on the outside, the [greatly enlarged] posterior end making the inside considerably curved; ligament pit very short and wide, width almost equalling length, and deepening as it approaches inside of hinge-plate where it breaks through, resembling in this respect the European species *P. vincentianum*.

Characterized by its very small size, triangular shape, heavy shell, and strangely shaped ridges on the beaks. Unlike any other species.

DISTRIBUTION
Known only from a few mostly disjunct localities in Ontario (Thames River) and the central United States south to Arkansas and Alabama.

ECOLOGY
Very rare. Reported to live in mud among dead leaves and aquatic plants and to be so covered with a black or brown coat that it resembles a globule of soil. Its distribution implies that it occurs only in hard-water habitats. Nothing is known about its reproduction or anatomy.

177
Pisidium cruciatum
a,b: Honeoye Creek, Rush, N.Y. ($a \times 27$; $b \times 41$).
c,d: Raisin R. near Norwell, Mich. ($c \times 27$; $d \times 11$).
e–j: Grand R., Pottawattomie Bayou, Ottawa Co., Mich. ($\times 16$).

178
Pisidium (Neopisidium) insigne Gabb, 1868
Tiny Pea Clam

DESCRIPTION
Shell very small, up to about 2 mm long, relatively high (H/L 0.82-0.84), compressed (W/L 0.47-0.59), elliptical, and thin shelled. Beaks low, rounded, and close to centre. Dorsal margin long and gently curved, anterior end broad and roundly pointed, ventral margin long and very gently curved, and posterior margin roundly truncate. Hinge plate narrow, long, and openly curved. Surface finely striate (about 25 to 30 per mm). Periostracum brownish and (in many specimens) with foreign material adhering to it.

The hinge teeth, according to Herrington (1962), are as follows:

> Laterals long and slim, cusps rather sharp on top; cusp of A2 distal or on distal side of center, of P2 very distal (the distance from this cusp to the cardinals seems very great), of A1 distal or on distal side of center, of P1 distal; cardinals of right valve subcentral or nearer anterior cusps; C2 small, almost straight, parallel with inside of hinge-plate; C4 small, rather indistinct, slightly curved, almost parallel with C2, posterior end a little nearer inside of hinge-plate and a little more posterior than C2; C3 slim, straight, or slightly curved, frequently slightly enlarged at posterior end, which is a little nearer the inside of hinge-plate.

Characterized by its very small size, low and elongate form, and broad rounded anterior end. Compare with *P. conventus* and with juvenile *P. casertanum*.

DISTRIBUTION
Known in Canada only from Prince Edward Island, southern Ontario, and southern British Columbia. Also occurs across the northern United States and south in the Rocky Mountains to Arizona and New Mexico.

ECOLOGY
Found principally in slow-moving creeks and spring creeks. Almost nothing is known about its biology.

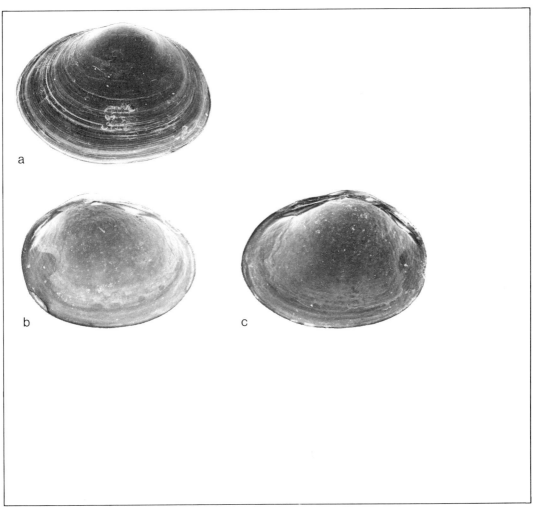

178
Pisidium insigne
a: Lidstone Creek, Prince Co., P.E.I. (× 22).
b,c: Long L., Wellington, B.C. (× 22).

179
Pisidium (Neopisidium) punctatum Sterki, 1895
Perforated Pea Clam

DESCRIPTION
Shell minute, up to about 1.7 mm long, relatively high (H/L 0.82-0.94), inflated (W/L 0.56-0.71), ovate, quite thin, and full of tiny holes that are clearly visible at magnification of 50×. In many specimens a transverse more or less concentric ridge is present near the umbones. Dorsal and ventral margins openly curved; anterior margin roundly pointed centrally and flattened above; posterior margin somewhat truncated, curved, and joining the dorsal margin with an angle. Hinge plate more than 3/4 shell length. In addition to the tiny holes, the surface is covered with rather coarse, concentric, evenly spaced striae. Periostracum very thin and pale yellowish brown.

The hinge teeth, according to Sterki (1895), are as follows:

> Hinge moderately strong; cardinal teeth fine, in the left valve two, lamellar, longitudinal, about equally long, a little curved, almost parallel, the upper little anterior; in the right valve one, longitudinal, little curved, lamellar, the posterior end slightly thickened; lateral teeth rather small and thin, in the left valve one, pointed, in the right valve two, the outer quite small.

Best distinguished by its very small size, regular but rather coarse striae, and tiny holes (or punctae) that cover the shell. The umbonal ridge, when present, is also distinctive. Closely related to *P. punctiferum* Guppy, but that species is much larger, more finely striate, and occurs only in the southern United States.

DISTRIBUTION
Found in scattered localities in southern Canada from Ontario to British Columbia, and in the northern United States.

ECOLOGY
Uncommon. Occurs in lakes and in slow-moving portions of rivers and streams. Found among vegetation and usually on muddy bottoms. Adults bear 2 litters of young each year, 1 in the spring and 1 in the fall.

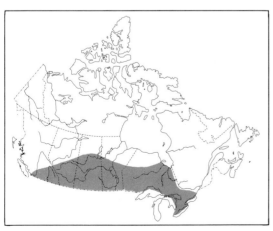

179
Pisidium punctatum
a,b,c: Athol Bay, L. Ontario, Prince Edward Co., Ont. (× 27).

Glossary

abaxial
outward, away from shell axis.

adaxial
inward, toward shell axis.

adductor muscle
a large muscle which closes the valves of a bivalved mollusc shell.

alate
expanded like a wing.

annuli (sing. annulus)
ringlike thickenings, generally related to previous growth pauses.

apex
top of the spire, the first formed part of a gastropod (snail) shell, pointed in most species.

arcuate
bent or arc-like.

beak
the earliest formed part of a bivalved shell. *See* umbo.

bifid
partly but not wholly split into two similar parts.

biconvex
convex on both sides.

calcareous
containing and/or resembling calcium carbonate.

callus
thickened shelly structure that covers inner lip or columellar region of a gastropod shell.

cardinal
central or major, as in cardinal hinge teeth.

carina
prominent spiral keel or ridge.

carinate
bearing one or more spiral keels or ridges.

collabral
conforming to shape of outer lip at an earlier growth stage as shown by growth lines.

columella
centrally located pillar surrounding the axis of coiling of a spiral gastropod shell.

conoid
shaped like a cone.

cusp
a projecting peak on a radular tooth.

decurrent
flowing downward.

dextral
right-handed or coiled in a clockwise direction when viewed from above; when viewed from the front (apex above) the aperture is on the right.

dimorphism
the condition of having two distinct forms, especially as in sexual dimorphism, in which males and females are visibly different.

dioecious
having the male reproductive system in one individual and the female in another.

disc
the rounded anterior portion of a freshwater mussel shell.

distal
far from the point of origin or attachment, opposite from proximal.

dorsal
the back, or, in bivalved molluscs, the area that includes the umbones and the ligament.

eutrophic lake
a shallow lake ordinarily characterized by an abundance of plant and animal life and a muddy bottom with high organic content. Oxygen concentrations are low in deeper water during the summer. This is a late stage in normal lake succession from the oligotrophic condition through mesotrophic to eutrophic, leading eventually to a marsh and finally to dry land.

globose
inflated and approaching the shape of a sphere.

glochidia (sing. glochidium)
the larvae of freshwater mussels (superfamily Unionacea).

gravid
pregnant, that is with fertilized eggs or unborn young within the body.

growth rest
a ridge formed during an intermediate stage of growth when this area was the edge of the shell.

helicone
a cone twisted into a spiral form; the distally expanded coiled tube that forms most gastropod shells.

hermaphrodite
an individual that has both male and female sexual organs.

hinge
the structure that joins the two halves of a bivalve shell at the dorsal margin. It is generally composed of an elastic ligament and articulating hinge teeth.

hinge plate
the thickened edge of a bivalved mollusc shell to which hinge teeth (if present) are attached.

immersed
depressed below the adjacent whorls, as an immersed nuclear whorl.

impressed
lying below the general adjacent surface as if stamped into it.

inflated
expanded and distended.

interdentum
in freshwater mussels the area of the hinge plate between the pseudocardinal and the lateral hinge teeth.

lamellae
thin plates or blade-like ridges.

lentic
characteristic of standing water, opposite to lotic.

lip
the structure surrounding, either entirely or in part, the mouth or aperture of a snail shell.

lotic
characteristic of running water, opposite to lentic.

lymnaeiform
shaped more or less like a typical *Lymnaea*, that is like *L. stagnalis*.

malleate
bearing flattened areas as if hammered.

mantle
the fleshy sheet of tissue that secretes the shell of a mollusc and is appressed to its inner surface. It encloses the mantle cavity and most, or all, of the internal organs.

marsupial
in molluscs refers to the possession of a structure for containing or enclosing the young.

mesotrophic lake
a lake of moderate depth with intermediate abundance of plants and animals. *See* oligotrophic lake and eutrophic lake.

monoecious
having both the male and female reproductive systems in the same individual.

morph
a distinct form or colour phase of a variable species.

nacre
the pearly inner layer characteristic of some mollusc shells.

nuclear whorl
the first turn of a spiral shell, beginning at the apex and corresponding to part or all of the larval shell.

obtuse
an angle exceeding 90°.

oligotrophic lake
a deep lake with sparse plant and animal life, and with low organic content in bottom deposits. Oxygen concentrations in such a lake remain high in deep water throughout the year. This is the first stage in normal lake succession. *See* eutrophic lake and mesotrophic lake.

operculum
a horny or shelly plate-like structure attached to the foot of most prosobranch gastropods. It seals the shell aperture when the animal is contracted within the shell.

ovate
oval; used also in combinations such as ovate-quadrate, meaning intermediate between ovate and quadrate.

palatal lip
the outer lip of a snail shell, that is the portion of the lip on the outer side of the aperture.

pallial line
an impressed line on the inner surface of a bivalved mollusc shell that marks the zone of attachment of the mantle to the shell.

parietal
in gastropods, pertaining to that part of the aperture adjacent to, or pressed against, the preceding whorl.

parthenogenic
capable of reproduction by direct development, without fertilization of the eggs.

patulous
broadly expanded or limpet-like.

paucispiral
consisting of few spirals or of less than one complete spiral.

pelagic
capable of swimming, or floating, in open water for extended periods.

penultimate
next to the last. The penultimate whorl is the one just above the body whorl.

peristome
lip.

periostracum
the outer parchment-like layer of a mollusc shell.

plait
a spiral, flattened ridge on the columella.

planorboid
coiled approximately in a single plane or resembling typical species of the family Planorbidae.

planospiral
coiled in a single plane.

plicae
broad ridges.

post-basally
located in the basal (ventral) region near the posterior end.

prosocline
with plane of aperture inclined away from the axis in its upper part, and toward the axis in its lower part. This is characteristic of most prosobranch gastropods.

protoconch
the larval shell of a mollusc. It is still visible in some partly mature and adult specimens.

pseudocardinal
the centrally located, short (or stump-like) hinge teeth of most bivalve molluscs.

punctate
the surface pitted with tiny holes.

quadrate
more-or-less four-sided, resembling a square or rectangle.

scalariform
loosely coiled with whorls not touching at the sides or tending toward this condition. See illustration of *Valvata sincera ontariensis*.

sculpture
impressed or raised markings.

sensu lato
in the broad sense (abbreviated *s. lat.*)

sensu stricto
in the strict sense (abbreviated *s. str.*)

septate
with one or more internal shelly partitions.

serrate
with a series of notches or grooves at the edge, like a saw or a file.

shouldered
shaped like a shoulder, with a flattened upper surface bounded by a definite angle.

sigmoid
shaped like an S.

sinistral
left-handed or coiled in a counterclockwise direction when viewed apically, that is from above. Viewed from the front (apex above) the aperture is on the left.

sinuate
wavy or sinuous.

spire
the upper surface of a spiral snail shell. In most species it is cone-shaped and tapers to a point (the apex). The last whorl (the body whorl) is not considered to be part of the spire. In flatly coiled snails, especially ramshorn snails (Planorbidae), the spire may be flat or concave but is recognizable because it is less concave than the base of the shell that contains the umbilicus.

spire angle
the angle at the apex formed by the cone of the spire.

sp.
abbreviation for species (singular).

spp.
abbreviation for species (plural).

ssp.
abbreviation for subspecies (singular).

sspp.
abbreviation for subspecies (plural).

striae
impressed lines or narrow grooves.

subglobose
inflated but not so much as to be round or globose.

substrate
the substance on which an organism lives or grows, such as soil, gravel, or rocks.

suffusion
an overspreading, as with a tint of colour.

suture
a spiral line or groove marking the junction of adjacent whorls.

tabulate
shaped like a table.

trapezoidal
shaped like a trapezoid, that is a four-sided geometric figure similar to a rectangle but with only two parallel sides.

truncate
shaped as if cut off.

umbilicus
the hollow centre (if present) of the axis of rotation of a snail shell, visible from the base of the shell.

umbo
the apex, or juvenile shell, of one half (a valve) of a bivalve mollusc.

umbones
plural of umbo.

varix
a prominent ridge or protrusion formed by expansion of the shell aperture during a previous stage of growth.

ventral
the underside or lower part of the shell. In bivalves it is on the opposite side from the beaks and hinge.

verge
in some gastropods an organ of the male genital tract that bears the penis.

vernal
occurring in the spring, for example a vernal pool.

whorl
a single complete turn of a spiral shell.

References

Serious students are referred to the following literature for additional information. The list includes not only the references cited elsewhere in this book by author and date only, but also the important works published in this field since 1968. For more extensive bibliographies of Canadian freshwater molluscs, consult especially Clarke (1973), LaRocque (1953), Taylor (1975), and the section on Mollusca in the *Zoological Record*, published annually (since 1864) by the Zoological Society of London.

Abbott, R.T.
(1974). American Seashells: The Marine Mollusca of the Atlantic and Pacific Coasts of North America. 2nd ed. New York: Van Nostrand Reinhold. 663 pp.

Bailey, R.M.; Fitch, J.E.; Herald, E.S.; Lachner, E.A.; Lindsey, C.C.; Robins, C.R.; and Scott, W.B.
(1970). A List of Common and Scientific Names of Fishes from the United States and Canada. 3rd ed. American Fisheries Society, Special Publication 6. 150 pp.

Baker, F.C.
(1928*a*). The Fresh-Water Mollusca of Wisconsin, Part I: Gastropoda. Wisconsin Geological and Natural History Survey, Bulletin 70. 507 pp.
(1928*b*). The Fresh-Water Mollusca of Wisconsin, Part II: Pelecypoda. Wisconsin Geological and Natural History Survey, Bulletin 70. 495 pp.

Baker, H.B.
(1925). Anatomy of *Lanx*, a Limpet-like Lymnaeid Mollusk. Proceedings of the California Academy of Sciences, Series 4, 14(8): 143-69.

Basch, P.F.
(1963). A Review of the Recent Freshwater Limpet Snails of North America (Mollusca: Pulmonata). Harvard University, Museum of Comparative Zoology Bulletin 129(8): 401-61.

Berry, E.G.
(1943). The Amnicolidae of Michigan: Distribution, Ecology, and Taxonomy. Miscellaneous Publications, Museum of Zoology, University of Michigan, No. 57. 68 pp.

Bousfield, E.L.
(1960). Canadian Atlantic Sea Shells. Ottawa: Department of Northern Affairs and National Resources, National Museum of Canada. 72 pp.

Burch, J.B.
(1962). How to Know the Eastern Land Snails. Dubuque, Iowa: Brown. 214 pp.
(1975*a*) Freshwater Sphaeriacean Clams (Mollusca: Pelecypoda) of North America. Rev. ed. Hamburg, Mich.: Malacological Publications. 96 pp. (Previously published in 1972 by the U.S. Environmental Protection Agency in Biota of Freshwater Ecosystems, Identification Manual No. 3. Washington, D.C.: U.S. Government Printing Office. 31 pp.)
(1975*b*) Freshwater Unionacean Clams (Mollusca: Pelecypoda) of North America. Rev. ed. Hamburg, Mich.: Malacological Publications. 204 pp. (Previously published in 1973 by the U.S. Environmental Protection Agency in Biota of Freshwater Ecosystems, Identification Manual No. 11. Washington, D.C.: U.S. Government Printing Office. 176 pp.)

Clampitt, P.T.
(1970). Comparative Ecology of the Snails *Physa gyrina* and *Physa integra* (Basommatophora: Physidae). Malacologia 10(1): 113-51.

Clarke, A.H.
(1973). The Freshwater Molluscs of the Canadian Interior Basin. Malacologia 13(1-2). 509 pp.
(1976). Endangered Freshwater Mollusks of Northwestern North America. Bulletin of the American Malacological Union Inc. for 1976: 18-19.

Clarke, A.H., and Berg, C.O.
(1959). The Freshwater Mussels of Central New York, with an Illustrated Key to the Species of Northeastern North America. Cornell University, Agricultural Experiment Station, Memoir 367. 79 pp.

Dazo, B.C.
(1965). The Morphology and Natural History of *Pleurocera acuta* and *Goniobasis livescens* (Gastropoda: Cerithiacea: Pleuroceridae). Malacologia 3(1). 80 pp.

Dundee, D.S.
(1957). Aspects of the Biology of *Pomatiopsis lapidaria* (Say) (Mollusca: Gastropoda: Prosobranchia). Miscellaneous Publications, Museum of Zoology, University of Michigan, No. 100. 37 pp.

Goodrich, Calvin
(1942). The Pleuroceridae of the Pacific Coastal Drainage, including the Western Interior Basin. Occasional Papers of the Museum of Zoology, University of Michigan, No. 469. 4 pp.
(1945). *Goniobasis livescens* of Michigan. Miscellaneous Publications, Museum of Zoology, University of Michigan, No. 64. 36 pp.

Haas, Fritz
(1969). Superfamilia Unionacea. Das Tierreich 88. 663 pp.

Hanna, G.D.
(1966). Introduced Molluscs of Western North America. Occasional Papers, California Academy of Sciences, No. 48, 108 pp.

Hart, C.W., Jr., and Fuller, S.L.H. (eds.)
(1974). Pollution Ecology of Freshwater Invertebrates. New York and London: Academic Press. 389 pp.

Heard, W.H.
(1965). Comparative Life Histories of North American Pill Clams (Sphaeriidae: *Pisidium*). Malacologia 2(3): 381-411.
(1975). Sexuality and other aspects of reproduction in *Anodonta* (Pelecypoda: Unionidae). Malacologia 15(1): 81-103.

Heard, W.H., and Guckert, R.H.
(1970). A Re-evaluation of the Recent Unionacea (Pelecypoda) of North America. Malacologia 10(2): 333-55.

Herrington, H.B.
(1962). A revision of the Sphaeriidae of North America (Mollusca: Pelecypoda). Miscellaneous Publications, Museum of Zoology, University of Michigan, No. 118. 74 pp.

Hubendick, Bengt
(1951). Recent Lymnaeidae, Their Variation, Morphology, Taxonomy, Nomenclature, and Distribution. Kunglica Svenska Vetenskapsakademiens Handlingar, Fjärde Serien, 3(1). 223 pp.
(1964). Studies on Ancylidae, the subgroups. Göteborgs Kungliga Vetenskaps och Vitterhets-Samhälles Handlingar, Sjätte Följden, Series B, 9(6): 1-72.

Johnson, R.I.
(1970). The Systematics and Zoogeography of the Unionidae (Mollusca: Bivalvia) of the Southern Atlantic Slope Region. Harvard University, Museum of Comparative Zoology Bulletin 140(6): 263-449.

Kuiper, J.G.J.
(1975). Zwei neue boreale *Pisidium*-Arten: *P. hinzi* und *P. waldeni*. Archiv für Molluskenkunde 106(1/3): 27-37.

LaRocque, Aurèle
(1953). Catalogue of the Recent Mollusca of Canada. National Museum of Canada Bulletin 129 (Biological Series 44). 406 pp.

Mackie, G.L.
(1973). Biology of *Musculium securis* (Pelecypoda: Sphaeriidae). . . . Ph.D. dissertation, University of Ottawa. 175 pp.

Mayr, Ernst
(1969). Principles of Systematic Zoology. New York: McGraw-Hill. 428 pp.

Moore, R.C. (ed.)
(1969). Treatise on Invertebrate Paleontology, Part N. Mollusca 6, Bivalvia. Vol. 2 of 3. New York and Lawrence, Kans.: Geological Society of America and University of Kansas. N491-N951.

Morrison, J.P.E.
(1955). Notes on the Genera *Lanx* and *Fisherola* (Pulmonata). Nautilus 68(3): 79-83.

Odhner, N.H.
(1929). Die Molluskenfauna des Tåkern. Sjön Tåkerns Fauna och Flora, utgiven av K. Svenska Vetenskapsakademien. Stockholm: Almqvist and Wiksells. 129 pp.

Pilsbry, H.A.
(1925). The Family Lancidae Distinguished from the Ancylidae. Nautilus 38(3): 73-75.
(1939-48). Land Mollusca of North America (North of Mexico). Monographs of the Academy of Natural Sciences of Philadelphia, No. 3, 1(1): 1-573 (1939); 1(2): 575-994 (1940); 2(1): 1-520 (1946); 2(2): 521-1113 (1948).

Ross, H.H.
(1974). Biological Systematics. Reading, Mass.: Addison-Wesley Publishing Co. 345 pp.

Sinclair, R.M., and Isom, B.G.
(1963). Further Studies on the Introduced Asiatic Clam (*Corbicula*) in Tennessee. Tennessee Department of Public Health, Tennessee Stream Pollution Control Board. 76 pp.

Solem, G.A.
(1974). The Shell Makers: Introducing Mollusks. Toronto: John Wiley and Sons. 289 pp.

Sterki, V.
(1895). Two New Pisidia. Nautilus 8(9): 97-100.

Taylor, D.W.
(1966). Summary of North American Blancan Non-Marine Mollusks. Malacologia 4(1): 1-172.
(1975). Index and Bibliography of Late Cenozoic Freshwater Mollusca of Western North America. Claude W. Hibbard Memorial Vol. 1. Museum of Paleontology Papers on Paleontology, No. 10. 384 pp.

Taylor, D.W., and Sohl, N.F.
(1962). An Outline of Gastropod Classification. Malacologia 1(1): 7-32.

Taylor, D.W.; Walter, H.J.; and Burch, J.B.
(1963). Freshwater Snails of the Sub-genus *Hinkleyia* (Lymnaeidae: *Stagnicola*) from the Western United States. Malacologia 1(2): 237-81.

Te, G.A.
(1975). Michigan Physidae, with Systematic Notes on *Physella* and *Physodon* (Basommatophora: Pulmonata). Malacological Review 8 (1/2) 7-30.

Van der Schalie, H., and Dundee, D.S.
(1955). The Distribution, Ecology and Life History of *Pomatiopsis cincinnatiensis* (Lea), an Amphibious Operculate Snail. Transactions of the American Microscopical Society 74(2): 119-33.

Index to Scientific and Common Names

References are to the consecutive numbers assigned to the species in the text. Italic numbers indicate that the taxon is briefly discussed under that species entry. Family names are listed in the table of contents.

Acella haldemani 36
Acroloxus coloradensis 23
Actinonaias carinata 131
acuta, Pleurocera 20
adamsi, Pisidium 160
Adam's Pea Clam 160
Alaskan Pond Snail 40
Alasmidonta
   calceola (= *Alasmidonta viridis*) 102
   heterodon 103
   marginata 104
   undulata 105
   varicosa 106
   viridis 102
alata, Proptera 126
alberta, Bakerilymnaea bulimoides morph *31*
Alewife Floater 118
ambigua, Simpsoniconcha 110
Amblema plicata 94
American Ear Snail 35
Amnicola
   limosa 14
   walkeri 15
amnicum, Pisidium 157
Amphibious Fossaria 29
anceps, Helisoma anceps 77
angulata, Gonidea 93
Anodonta
   beringiana 112
   cataracta cataracta 113
   cataracta fragilis 114
   grandis grandis 115
   grandis simpsoniana 116
   imbecilis 117
   implicata 118
   kennerlyi 119
   nuttalliana 120
   oregonensis (= *Anodonta nuttalliana*) 120
   wahlamatensis (= *Anodonta nuttalliana*) 120
Anodontoides ferussacianus 111
Aplexa hypnorum 65
appressa, Lymnaea stagnalis (= *Lymnaea stagnalis jugularis*) 38
arctica, Stagnicola 43
Arctic-Alpine Fingernail Clam 147
Arctic-Alpine Pea Clam 176
armigera, Planorbula 75
Armiger crista 70
Asiatic Clam 144
athearni, Physa jennessi 56
atkaensis, Lymnaea 40
auricularia, Radix 33

Bakerilymnaea
   bulimoides 31
   bulimoides morph *alberta* *31*
     morph *bulimoides* 31
     morph *cockerelli* 31
     morph *perplexa* 31
     morph *techella* 31
     morph *vancouverensis* 31
   dalli 32
Banded Mystery Snail 2
Banff Springs Physa 58
Bean Villosa 140
Bell-mouthed Ramshorn 79
beringiana, Anodonta 112
binneyi, Helisoma trivolvis 85
Binney's Stout Ramshorn 85
Bithynia tentaculata 19
Black Sand-Shell 133
Blade-ridged Stagnicola 41
Blunt Albino Physa 56
Blunt Arctic Physa 55
Blunt Prairie Physa 57
British Columbia Menetus 74
Broad Promenetus 72
Brook Lasmigona 108
Brook Wedge Mussel 102
Brown Mystery Snail 1
Bulimnea megasoma 37
bulimoides, Bakerilymnaea 31
bulimoides, Bakerilymnaea bulimoides morph 31
Bulimus tentaculatus (= *Bithynia tentaculata*) 19
Bythinia tentaculata (= *Bithynia tentaculata*) 19

calceola, Alasmidonta (= *Alasmidonta viridis*) 102
campanulatum, Helisoma campanulatum 79
Campeloma
   decisum 1
   integrum *1*
Campeloma Spire Snail 10
campestris, Planorbula 76
Capacious Manitoba Ramshorn 81
caperata, Stagnicola 41
carinata, Actinonaias 131
cariosa, Lampsilis 134
Carunculina parva 127
casertanum, Pisidium 161
cataracta, Anodonta cataracta 113
catascopium, Stagnicola catascopium 44
chinensis, Cipangopaludina 3
Cincinnatia cincinnatiensis 10
cincinnatiensis, Cincinnatia 10
cincinnatiensis, Pomatiopsis 18
Cipangopaludina chinensis 3
circumstriatus, Gyraulus 66
coccineum, Pleurobema 101
cockerelli, Bakerilymnaea bulimoides morph 31

collinsi, Helisoma campanulatum   80
coloradensis, Acroloxus   23
columbiana, Physa   59
Columbia River Physa   59
columella, Pseudosuccinea   35
Common Floater   115
Common Stagnicola   47
complanata, Elliptio   99
complanata, Lasmigona   107
compressa, Lasmigona   108
compressum, Pisidium   162
concolor, Physa   60
contectoides, Viviparus (= *Viviparus georgianus*)   2
contortum, Pisidium nitidum form   169
conventus, Pisidium   176
cooperi, Menetus   74
Corbicula
   fluminea   144
   leana (= *Corbicula fluminea*)   144
   manilensis (= *Corbicula fluminea*)   144
corneum, Sphaerium   145
corpulentum, Helisoma corpulentum   81
costata, Lasmigona   109
crista, Armiger   70
cristatum, Pisidium lilljeborgi morph   167
cruciatum, Pisidium   177
Cyclonaias tuberculata   98
Cylindrical Floater   111

dalli, Bakerilymnaea   32
decampi, Fossaria   25
decepta, Marstonia   12
decisum, Campeloma   1
Deep-Water Spire Snail   16
Deer-Toe   125
deflectus, Gyraulus   67
Delicate Lamp-Mussel   136
dilatata, Elliptio   100
donaciformis, Truncilla   124
dubium, Pisidium   158
Dusky Lily-Pad Limpet   87
Dwarf Wedge Mussel   103
Dysnomia
   torulosa rangiana   142
   triquetra   143

Eastern Elliptio   99
Eastern Floater   113
Eastern Lamp-Mussel   137
Eastern Physa   53
Eastern-River Pearl Mussel   91
Elliptio
   complanata   99
   dilatata   100
elodes, Stagnicola   47
equilaterale, Pisidium   163
European Ear Snail   33
European Fingernail Clam   145
European Valve Snail   5
exacuous, Promenetus exacuous   71
exigua, Fossaria   26

fabale, Sphaerium   146
fabalis, Villosa   140
falcata, Margaritifera   92
fallax, Pisidium   164
False Pig-Toe   101
fasciola, Lampsilis   135
fasciolaris, Ptychobranchus   122
Fat Mucket   138
Fat Pea Clam   170
Faucet Snail   19
Fawn's-Foot   124
Ferrissia
   fragilis   88
   fragilis form *isabellae*   88
   parallela   89
   rivularis   90
ferruginea, Fossaria   27
ferrugineum, Pisidium   165
ferussacianus, Anodontoides   111
Flat-ended Spire Snail   11
Flatly Coiled Gyraulus   66
Flat-sided Horn Snail   20
Flat-sided Lake Limpet   89
Flat Valve-Snail   4
flava, Fusconaia   95
Flumincola virens (= *Lithoglyphus virens*)   17
fluminea, Corbicula   144
Fluted Shell   109
Fossaria
   decampi   25
   exigua   26
   ferruginea   27
   modicella   28
   modicella morph *rustica*   28
   parva   29
   truncatula   30
Fragile Fossaria   27
Fragile Paper-Shell   130
fragilis, Anodonta cataracta   114
fragilis, Ferrissia   88
fragilis, Leptodea   130
Fusconaia flava   95
fuscus, Laevapex   87

Gatineau Tadpole Snail   52
georgianus, Viviparus   2
Giant Columbia-River Limpet   24
Giant Columbia-River Spire Snail   17
Giant Fingernail Clam   150
Giant Manitoba Ramshorn   81
Giant Northern Pea Clam   159
Giant Promenetus   72
Giant Western Physa   62
Giant Western Spire Snail   17
Globular Pea Clam   174
Gonidea angulata   93
Goniobasis livescens   21
Graceful Fossaria   26
Graceful Keeled Horn Snail   22
grandis, Anodonta grandis   115
granum, Lyogyrus   13

Greater Carinate Ramshorn   83
Greater Columbia-River Limpet   24
Greater Eastern Pea Clam   158
Greater European Pea Clam   157
Great Lakes Horn Snail   21
Great Pond Snail   38
Grooved Fingernail Clam   150
Gyraulus
   circumstriatus   66
   deflectus   67
   parvus   68
   vermicularis   69
gyrina, Physa gyrina   51

haldemani, Acella   36
Haldeman's Physa   60
Heavy-toothed Wedge Mussel   105
helicoidea, Valvata sincera   7
Helisoma
   anceps anceps   77
   anceps royalense   78
   campanulatum campanulatum   79
   campanulatum collinsi   80
   corpulentum corpulentum   81
   corpulentum vermilionense   82
   corpulentum whiteavesi   82
   multivolvis   80
   pilsbryi infracarinatum   83
   trivolvis binneyi   85
   trivolvis subcrenatum   86
   trivolvis trivolvis   84
henslowanum, Pisidium   166
Henslow's Pea Clam   166
Herrington's Fingernail Clam   152
heterodon, Alasmidonta   103
heterostropha, Physa   53
hindsii, Lithoglyphus   17
hordacea, Physa   61
Hump-backed Pea Clam   172
Hydrobia nickliniana   18
hypnorum, Aplexa   65

idahoense, Pisidium   159
imbecilis, Anodonta   117
implicata, Anondonta   118
infracarinatum, Helisoma pilsbryi   83
insigne, Pisidium   178
integra, Physa   54
integrum, Campeloma   1
iris, Villosa   141
Irregular Gyraulus   67
isabellae, Ferrissia fragilis form   88

japonicus, Viviparus (= *Cipangopaludina chinensis*)   3
jenksii, Planorbula (= *Planorbula armigera*)   75
jennessi, Physa jennessi   55
johnsoni, Physa   58
Juga
   plicifera   22
   silicula (= *Juga plicifera*)   22
jugularis, Lymnaea stagnalis   38

Keeled Promenetus   71
kennerlyi, Anodonta   119
kennicotti, Stagnicola   48
Kidney Shell   122

lacustre, Sphaerium   153
lacustris, Probythinella   11
Lady-Finger   100
Laevapex fuscus   87
Lake Fingernail Clam   153
Lake Stagnicola   44
Lake Superior Ramshorn   78
Lampsilis
   cariosa   134
   fasciola   135
   ochracea   136
   ovata   139
   ovata ventricosa (= *Lampsilis ventricosa*)   139
   radiata radiata   137
   radiata siliquoidea   138
   ventricosa   139
Lanx nuttalli   24
lapidaria, Pomatiopsis   18
Larger Eastern Ramshorn   84
Larger Prairie Ramshorn   86
Lasmigona
   complanata   107
   compressa   108
   costata   109
latchfordi, Physa gyrina   52
leana, Corbicula (= *Corbicula fluminea*)   144
Leptodea fragilis   130
letsoni, Pyrgulopsis   12
Ligumia
   nasuta   132
   recta   133
Lilliput Mussel   127
lilljeborgi, Pisidium   167
Lilljeborg's Pea Clam   167
limosa, Amnicola   14
Lithoglyphus
   hindsii   17
   virens   17
Liver-Fluke Fossaria   30
livescens, Goniobasis   21
Long Fingernail-Clam   156
Loosely Coiled Valve Snail   8
lordi, Physa   62
Low-spired Ramshorn   80
Lymnaea
   atkaensis   40
   palustris (= *Stagnicola elodes*)   47
   peregra (= *Radix peregra*)   34
   stagnalis appressa (= *Lymnaea stagnalis jugularis*)   38
   stagnalis jugularis   38
   stagnalis sanctaemariae   39
   stagnalis wasatchensis (= *Lymnaea stagnalis jugularis*)   38
Lyogyrus granum   13

macrodon, Truncilla 124
mainense, Pisidium walkeri form 175
malleatus, Viviparus (= *Cipangopaludina chinensis*) 3
manilensis, Corbicula (= *Corbicula fluminea*) 144
Maple-Leaf 96
Margaritifera
  falcata 92
  margaritifera 91
margaritifera, Margaritifera 91
marginata, Alasmidonta 104
Marstonia decepta 12
megas, Promenetus exacuous 72
megasoma, Bulimnea 37
Menetus cooperi 74
mergella, Valvata 6
milium, Pisidium 168
Miniature Lake-Stagnicola 45
Modest Fossaria 28
Modest Gyraulus 68
modicella, Fossaria 28
montanensis, Stagnicola 42
Mountain-Spring Stagnicola 42
Mucket 131
Mudpuppy Mussel 110
multivolvis, Helisoma 80
Muskeg Stagnicola 43

nasoni, Stagnicola catascopium 45
nasuta, Ligumia 132
Newfoundland Floater 114
nickliniana, Hydrobia 18
nitidum, Pisidium 169
nitidum, Sphaerium 147
Northern Floater 116
Northern Riffle Shell 142
Northern Valve Snail 7
nuttalli, Lanx 24
nuttalli, Physa 63
nuttalliana, Anodonta 120
Nuttall's Physa 63

Obliquaria reflexa 123
Obovaria
  olivaria 128
  subrotunda 129
occidentale, Sphaerium 152
ochracea, Lampsilis 136
olivaria, Obovaria 128
Olive Hickory-Nut 128
ontariensis, Valvata sincera 8
Ordinary Spire Snail 14
oregonensis, Anodonta (= *Anodonta nuttalliana*) 120
Oriental Mystery Snail 3
Ornamented Pea Clam 177
Oval Lake-Limpet 88
ovata, Lampsilis 139

Pacific Coast Gyraulus 69
palustris, Lymnaea (= *Stagnicola elodes*) 47
palustris, Stagnicola (= *Stagnicola elodes*) 47
Paper Pond-Shell 117
parallela, Ferrissia 89
partumeium, Sphaerium 154
parva, Carunculina 127
parva, Fossaria 29
parvus, Gyraulus 68
patella, Sphaerium 148
pauperculum, Pisidium nitidum form 169
perdepressa, Valvata 4
peregra, Radix 34
Perforated Pea Clam 179
perplexa, Bakerilymnaea bulimoides morph 31
Physa
  columbiana 59
  concolor 60
  gyrina gyrina 51
  gyrina latchfordi 52
  heterostropha 53
  hordacea 61
  integra 54
  jennessi athearni 56
  jennessi jennessi 55
  jennessi skinneri 57
  johnsoni 58
  lordi 62
  nuttalli 63
  propinqua 64
  vinosa 51
  virginea 62
Pig-Toe 95
Pilsbry's Spire Snail 12
Pink Heel-Splitter 126
piscinalis, Valvata 5
Pisidium
  adamsi 160
  amnicum 157
  casertanum 161
  compressum 162
  conventus 176
  cruciatum 177
  dubium 158
  equilaterale 163
  fallax 164
  ferrugineum 165
  henslowanum 166
  idahoense 159
  insigne 178
  lilljeborgi 167
  lilljeborgi morph *cristatum* 167
  milium 168
  nitidum 169
  nitidum form *contortum* 169
  nitidum form *pauperculum* 169
  punctatum 179
  punctiferum 179
  rotundatum 170
  subtruncatum 171
  supinum 172

variabile 173
ventricosum 174
vincentianum 177
waldeni 167
walkeri 175
walkeri form *mainense* 175
Planorbula
  armigera 75
  campestris 76
  jenksii (= *Planorbula armigera*) 75
Pleurobema coccineum 101
Pleurocera acuta 20
plicata, Amblema 94
plicifera, Juga 22
Pocket-Book 139
Pointed Lake Limpet 23
Pointed Sand-Shell 132
Polished Tadpole Snail 65
Pomatiopsis
  cincinnatiensis 18
  lapidaria 18
Pond Fingernail Clam 155
Prairie Pond Snail 31
Prairie Toothed Planorbid 76
preblei, Stagnicola catascopium 46
Probythinella lacustris 11
Promenetus
  exacuous exacuous 71
  exacuous megas 72
  umbilicatellus 73
propinqua, Physa 64
Proptera alata 126
proxima, Stagnicola 49
Pseudosuccinea columella 35
Ptychobranchus fasciolaris 122
punctatum, Pisidium 179
punctiferum, Pisidium 179
Purple Pimple-Back 98
pustulosa, Quadrula 97
Pyrgulopsis letsoni 12

Quadrangular Pill Clam 168
Quadrula
  pustulosa 97
  quadrula 96
quadrula, Quadrula 96

radiata, Lampsilis radiata 137
Radix
  auricularia 33
  peregra 34
Rainbow Shell 141
rangiana, Dysnomia torulosa 142
recta, Ligumia 133
reflexa, Obliquaria 123
reflexa, Stagnicola 50
rhomboideum, Sphaerium 149
Rhomboid Fingernail Clam 149
Ribbed Valve-Snail 6
Ridged-Beak Pea Clam 162
Ridged Wedge-Mussel 104

River-Bank Looping Snail 18
River Fingernail Clam 146
River Pea Clam 164
rivularis, Ferrissia 90
Rocky Mountain Fingernail Clam 148
Rocky Mountain Ridged Mussel 93
Rocky Mountain Stagnicola 49
rotundatum, Pisidium 170
Round Hickory-Nut 129
Round Pea Clam 163
rowelli, Stagnicola proxima (= *Stagnicola proxima*) 49
royalense, Helisoma anceps 78
rustica, Fossaria modicella morph 28
Rusty Pea Clam 165
Rusty Spire Snail 13

sanctaemariae, Lymnaea stagnalis 39
Say's Toothed Planorbid 75
securis, Sphaerium 155
Shiny Pea Clam 169
Short-ended Pea Clam 171
Shouldered Northern Fossaria 25
Showy Pond Snail 37
silicula, Juga (= *Juga plicifera*) 22
siliquoidea, Lampsilis radiata 138
simile, Sphaerium 150
simpsoniana, Anodonta grandis 116
Simpsoniconcha ambigua 110
sincera, Valvata sincera 6
skinneri, Physa jennessi 57
Slender Pond Snail 36
Small Pond-Snail 32
Small Spire-Snail 15
Solid Lake-Physa 54
Somatogyrus subglobosus 16
Sphaerium
  corneum 145
  fabale 146
  lacustre 153
  nitidum 147
  occidentale 152
  partumeium 154
  patella 148
  rhomboideum 149
  securis 155
  simile 150
  striatinum 151
  transversum 156
Spike 100
Squaw-Foot 121
Stagnicola
  arctica 43
  caperata 41
  catascopium catascopium 44
  catascopium nasoni 45
  catascopium preblei 46
  elodes 47
  kennicotti 48
  montanensis 42
  palustris (= *Stagnicola elodes*) 47
  proxima 49

445

proxima rowelli (= *Stagnicola proxima*) 49
   reflexa 50
   yukonensis (= *Stagnicola arctica*) 43
Striated Fingernail Clam 151
striatinum, Sphaerium 151
Striped Stagnicola 50
Strophitus undulatus 121
Sturdy River Limpet 90
Subarctic Lake-Stagnicola 46
subcrenatum, Helisoma trivolvis 86
subglobosus, Somatogyrus 16
subrotunda, Obovaria 129
subtruncatum, Pisidium 171
supinum, Pisidium 172
Swamp Fingernail Clam 154
Swollen Wedge-Mussel 106

Tadpole Snail 51
techella, Bakerilymnaea bulimoides morph 31
tentaculata, Bithynia 19
tentaculata, Bythinia (= *Bithynia tentaculata*) 19
tentaculatus, Bulimus (= *Bithynia tentaculata*) 19
Three-horned Warty-Back 123
Three-keeled Valve Snail 9
Three-Ridge 94
Tiny Nautilus Snail 70
Tiny Pea Clam 178
transversum, Sphaerium 156
Triangular Pea Clam 173
tricarinata, Valvata 9
Tricorn Pearly Mussel 143
triquetra, Dysnomia 143
trivolvis, Helisoma trivolvis 84
truncata, Truncilla 125
truncatula, Fossaria 30
Truncilla
   donaciformis 124
   macrodon 124
   truncata 125
tuberculata, Cyclonaias 98
Two-ridged Ramshorn 77

Ubiquitous Pea Clam 161
umbilicatellus, Promenetus 73
Umbilicate Promenetus 73
undulata, Alasmidonta 105
undulatus, Strophitus 121

Valvata
   mergella 6
   perdepressa 4
   piscinalis 5
   sincera helicoidea 7
   sincera ontariensis 8
   sincera sincera 6
   tricarinata 9
   virens 6
vancouverensis, Bakerilymnaea bulimoides morph 31
Vancouver Island Physa 61
variabile, Pisidium 173
varicosa, Alasmidonta 106

ventricosa, Lampsilis 139
ventricosa, Lampsilis ovata (= *Lampsilis ventricosa*) 139
ventricosum, Pisidium 174
vermicularis, Gyraulus 69
vermilionense, Helisoma corpulentum 82
Villosa
   fabalis 140
   iris 141
vincentianum, Pisidium 177
vinosa, Physa 51
virens, Lithoglyphus 17
virens, Valvata 6
virginea, Physa 62
viridis, Alasmidonta 102
Viviparus
   contectoides (= *Viviparus georgianus*) 2
   georgianus 2
   japonicus (= *Cipangopaludina chinensis*) 3
   malleatus (= *Cipangopaludina chinensis*) 3
   viviparus 2
viviparus, Viviparus 2

wahlamatensis, Anodonta (= *Anodonta nuttalliana*) 120
waldeni, Pisidium 167
walkeri, Amnicola 15
walkeri, Pisidium 175
Walker's Pea Clam 175
Walker's Pond Snail 39
Wandering Snail 34
Warty-Back 97
wasatchensis, Lymnaea stagnalis (= *Lymnaea stagnalis jugularis*) 38
Wavy-rayed Lamp-Mussel 135
Western Arctic Stagnicola 48
Western Floater 119
Western Lake Physa 64
Western-River Pearl Mussel 92
whiteavesi, Helisoma corpulentum 82
Whiteaves's Capacious Ramshorn 82
White Heel-Splitter 107
Winged Floater 120

Yellow Lamp-Mussel 134
yukonensis, Stagnicola (= *Stagnicola arctica*) 43
Yukon Floater 112